T0142807

Multiple Criteria Analysis
in Strategic Siting Problems

Multiple Criteria Analysis in Strategic Siting Problems

by

Oleg I. Larichev
Russian Academy of Sciences, Moscow, Russia

and

David L. Olson
Texas A&M University, Texas, USA

Kluwer Academic Publishers
Boston/Dordrecht/London

Distributors for North, Central and South America:
Kluwer Academic Publishers
101 Philip Drive
Assinippi Park
Norwell, Massachusetts 02061 USA
Telephone (781) 871-6600
Fax (781) 871-6528
E-Mail < kluwer@wkap.com >

Distributors for all other countries:
Kluwer Academic Publishers Group
Distribution Centre
Post Office Box 322
3300 AH Dordrecht, THE NETHERLANDS
Telephone 31 78 6392 392
Fax 31 78 6546 474
E-Mail < orderdept@wkap.nl >

 Electronic Services < http://www.wkap.nl >

Library of Congress Cataloging-in-Publication Data

A C.I.P. Catalogue record for this book is available from the Library of Congress.

Table of Contents

1 INTRODUCTION

Facility Location Problems

The location problem has been with humans for all of their history. In the past, many rulers had the decision of locating their capital. Reasons for selecting various locations included central location, transportation benefits to foster trade, and defensibility. The development of industry involved location problems for production facilities and trade outlets. Obvious criteria for location of business facilities included profit impact. In the 19th century, there seemed to be a focus on the cost of transporting raw materials versus the cost of transporting goods to consumers. Location decisions were made considering all potential gains and expenses. Some judgment was required, because while most benefits and costs could be measured accurately, not all could be. Successful business practice depended on the sound judgment of the decision-maker in solving location problems.

Each of these enterprises produced some wastes. Finding a location to dispose of these wastes was not a difficult task. In less-enlightened times, governments resorted to fiat and land-condemnation to take the sites needed for disposal. In the 19th century, industry grew rapidly in Great Britain and elsewhere as mass production served expanding populations of consumers. The by-products of mass-production were often simply discarded in the most expeditious manner. There are still mountains in the United States with artificial facades created from the excess material discarded from mining activity.

We have developed the ability to create waste of lethal toxicity. Some of this waste is on a small but potentially terrifying scale, such as plutonium. Other forms involve massive quantities that can convert entire regions into wasteland, and turn entire seas into man-made bodies of dead water. Siting facilities to deal with this waste is one of the most difficult decisions we face as a society.

Some waste is generated as part of someone's plan. Other forms arise due to accident, such as oil-spills or chemical plant catastrophes. Location decisions for waste-related facilities are very important. Dangerous facilities have been constructed in isolated places for the most part in the past. However, with time, fewer places in the world are all that isolated. Furthermore, moving toxic material safely to or from wherever these sites are compounds the problem.

The problem of locating facilities has been much more difficult in the 20th century, and promises to be even more problematic in the 21st century. Locating enterprises is now a much more complex decision. Many more qualitative criteria need to be considered, such as the impact on the environment, the possibility of accidents and spills, the consequences of such accidents, and so forth. A reliable means of transforming accident consequences into concrete cost results has not been evident. There is a need for decision-makers to solve multicriteria location problems.

The 20th century also saw the development of many different types of location decisions. New enterprises rely on cost-efficient production systems. Cost remains an important criterion for successful business. The construction of facilities and/or the processes of producing end products involve high levels of uncertainty. Enterprise activities involve exposure to possible disasters. Each new accident is the coincidence of several causes each having a low probability taken separately. There is insufficient reliable statistical data to accurately predict possible accidents and their consequences. We can give many examples of these difficulties in locating facilities. These include nuclear power stations, liquid gas terminals, large chemical factories, and oil and gas pipelines. We can refer to these as hazardous facilities.

The importance of location decisions is evident. Kleindorfer and Kunreuther (1987) discussed release of toxic chemicals from a plant in Seveso, Italy in 1976, explosion of liquified petroleum gas stored at San Juan Ixhuatepec, Mexico in 1984, and pesticide release at a chemical plant in Bhopal, India in 1984. At Seveso, a cloud of toxic herbicide vapors was released that led to extensive contamination of soil and buildings in an area populated by hundreds of thousands of people north of Milan. In San Juan Ixhuatepec, a large refinery located in the center of the densely populated capital district caused a fire that burned for 36 hours, destroying at least 20 city blocks. The Bhopal accident released a highly volatile chemical used in pesticides into a densely populated area. Bhopal and the Chernobyl nuclear plant explosion in 1986 (Medvedev, 1990) were cited by Kleindorfer and Kunreuther (p. 98) as being the two most disastrous pollution-related accidents in terms of deaths caused. Those catastrophes demonstrate the importance of sound decision making in site location for potentially dangerous facilities.

A second problematic decision is the location of waste disposal facilities. In the past, little attention was given to this problem, because there was so much empty and unexplored space. However, late in the 20th century a consensus has evolved making such haphazard disposal of wastes unacceptable. It seems evident that even greater attention will be required for these decisions in the 21st century.

Specific Features of 20th Century Location Problems

Twentieth century location problems can be divided into those of locating enterprises and those of locating waste facilities. Both problems have the following typical features:

1. The multicriteria nature of location decisions
    Usually there is a need for decision-makers to consider more than mere profit impact in location decisions. Some criteria are easily measured. Many, however, are qualitative in nature, defying accurate measurement. For those criteria that are measurable, measures are in different units that are difficult to balance. The general value of each alternative must integrate each of these different estimates. This

requires some means of integrating different measures based on sound data.

2. The strategic nature of location decisions
    Choice of a facility location usually is followed by construction of a plant or waste facilities. Construction is often time-consuming. Therefore, the time between the making of a decision and its implementation can be great. This leads to detailed studies of possible alternative sites and methods in order to implement a rational decision process.

3. Uncertain and unknown factors associated with location decisions
    Typically, some of the information required for a location decision is missing due to incomplete understanding of technical and scientific aspects of a project. Sometimes, information about the consequences of a decision can be obtained only after many years of observation after construction. For example, waste sites can last great periods, thousands of years in the case of nuclear waste. There is not sufficient knowledge to anticipate what might happen at particular locations. Often facility location decisions must be made with incomplete information.

4. Public participation in decision making
    At one time, individual leaders of countries and industries could make individual decisions. That is not the case late in the 20th century, nor will it be the case in the 21st century.
    Public opinion has become critically important in such decisions. The importance of public values has been noted in a number of studies. De Montgolfier and Bertier (1978) studied the positions of several active groups in the choice of a major road, Brown (1984) studied public values in water control in central Arizona, Renn, et al. (1984) public views of energy policies, Grassin (1986) public views of high voltage line location, and Azis (1990) investigated public values in evaluating the impact of a highway project in Sumatra. Ways to elicit public preferences have been presented by Jones, et al. (1990, British energy policy), Keeney, et al. (1990, German energy policy), and Georgopoulou, et al. (1998, Greek energy policy).

While we realize that wastes need to be disposed of, none of us want to expose our families or ourselves to a toxic environment. Smith and

Desvousges (1986) discussed the origin of the term LULU (Locally Undesirable Land Uses), which seems to appropriately capture the nature of such facilities. Inhaber (1992) provided one of many discussions of the "not-in-my-backyard" syndrome. Even when those near a proposed site are in favor of a particular project, there is an interesting doughnut effect in public reaction (Easterling and Kunreuther, 1995). Those near the proposed site often support the location of the site in their vicinity (usually for economic reasons). Those in the next adjacent areas have been found to be quite strongly opposed to the location, with opposition tapering off as distance from the site increases.

This concern on the part of governmental decision-making bodies and local citizens has led to greater care in siting waste facilities. While arbitrary decisions were possible in the past (or decisions made considering only technical and/or economic issues), the political climate today demands that the opinions and preferences of more people be considered. This has led to more thorough analysis of location decisions, with consideration given to multiple criteria. We will see that multiple criteria analysis often results in broadening the search to find better solutions. In the 1950s, for instance, when coal-fired and petroleum-fired electricity generation generated unacceptable levels of pollution, nuclear energy was marketed as a way to produce electricity with far less pollution (and lower cost). Complex social systems have an interesting way of confounding those who would manipulate them. Nuclear energy was found to involve unacceptable levels of risk of radiation exposure (while turning out to be one of the most expensive sources after environmental protection standards were established).

The Goal and Organization of the Book

There are two major classes of multicriteria facility location problems. In the first class are problems of choosing territorial location. A typical problem statement for this class of problem is given in Cohon (1978): A territory is covered by a grid. Each cell of the grid is a possible alternative for facility location. The characteristics of each alternative site are estimated in terms of several criteria. The decision is to select the best grid cell after consideration of the estimated performance of each over the criteria being considered. Multiobjective mathematical programming is usually used for this class of problem (i.e., Cohon, et al., 1980). After Cohon's pioneering work, problems of this class have been discussed in many papers and books (see the surveys of Current, et al., 1990; Solomon and Haynes, 1984). This class of problem usually involves the generation of alternatives by territory. It is presupposed that it is possible to locate the facility at any cell within each territory. We call this class of problem one of facility location on a territory.

The second class of facility location problem is the multicriteria choice of the best location from a set of given alternatives. For this class of problem we do not have freedom of territorial choice. Because of technical or political limitations, the possible location is limited to a relatively small number of sites. The solution tools for this class of facility location problem are quite different from those of the first class. We call tools for the second class of problem decision aids, or preference-based analysis.

The problems of multicriteria facility location choice are central to this book. This emphasis is due to the presence of more and more difficult large-scale practical problems of location. There are different techniques used for solution of such problems, with a great deal of heated discussion about the positive and negative features of different multicriteria methods.

The goal of the book is to analyze different multicriteria techniques oriented on the solution of strategic facility location problems. We present descriptive and normative approaches to the problem, with practical cases and methodological discussions. We try to demonstrate the great diversity in multicriteria location problems. The most well-known decision techniques are presented with case solutions. Chapter 2 presents the descriptive model of multicriteria and multiparty location decisions developed by H. Kunreuther (Kunreuther, et al., 1983). This model demonstrates how location decisions are actually made in the world. Along with interesting research findings, Kunreuther and Linnerooth proposed a technique to describe the decision process involved in facility location decision.

Chapter 2 will also review normative tools for analysis of multiple criteria decision making. The starting point for these methods is cost-efficiency analysis. When multiple criteria are added to the analysis, it quickly becomes very difficult to convert all things of value into monetary terms. Pricing the value of a life, especially someone else's life, is highly problematic. Normative tools of analysis have been developed in many different forms. In the United States, multiattribute utility theory and analytic hierarchy process methods have been widely applied. In France, different approaches with different assumptions of value are considered in outranking methods. In Russia, there has been a focus on the psychological validity of decision methods and on the reliability of input information. Verbal decision analysis methods use yet another approach to model the tradeoffs generated by multiple criteria.

In Chapter 2, a framework for analysis considering multiple attributes applied by multiple parties is also described. Multicriteria methods are described at the operational level. An example location problem is presented, and the implementation of the different approaches are illustrated using the example problem.

The multicriteria tools covered in this book have been widely applied to public policy decision making. Cost benefit analysis was addressed in Kunreuther and Easterling (1990) and Walker, et al. (1994). Powell (1996) applied a weighting method. Multiattribute utility theory has been implemented by Anandalingam (1987), Keeney, et al., (1995), Kim, et al. (1998), and studies referenced in chapters 3 and 4 of this book. Analytic hierarchy process has been applied in many studies, including Barbarosoglu and Pinhas (1995). Barda, et al. (1990), Georgopoulou, et al. (1997), Karagiannidia and Moussiopoulos (1997), and Srinivasa Raju and Pillai (1999a) have all applied ELECTRE outranking methods. Related PROMETHEE studies include Mladineo, et al. (1987), Briggs, et al. (1990), Vuk, et al. (1991), Mladineo, et al. (1992), Abu-Taleb and Mareschal (1995), Al-Kloub, et al. (1997), and Al-Shemmeri, et al. (1997). Verbal Decision Analysis has been used in several practical problems, including Larichev, et al. (1987) and Larichev and Moshkovich (1997). Less common methods have also been applied, such as ORESTE (Delhaye, et al., 1991) and NAIADE (Haastrup, et al., 1998). Studies comparing selected methods include Roy and Bouyssou (1986), Flanders, et al. (1998), Toland, et al., (1998), Salminen, et al. (1998), and Srinivasa Raju and Pillai (1999a, 1999b). Georgopoulou, et al. (1998) reported a group decision support system dealing with renewable energy choices using the PROMETHEE system. Haastrup, et al. (1998) presented another decision support system for selection of an urban waste management site reflecting the views of different parties involved in the decision.

Chapters three through seven follow implementations of these multiple criteria methods to location decisions. The applications were selected on the basis of the availability of multiple reports of their use in order to demonstrate more completely how each method deals with multiple criteria, and the type of decision-making analysis they provide. Models are described in some detail, based on the original publications. We have described these applications using our own figures and tables following the originals in order to thoroughly examine the methods in question in some detail. We also follow the original applications in order to look at the context of group processes applied in real practice.

Chapter 3 will review one of the most thoroughly published applications of multiple criteria decision analysis to a selection problem. This decision involved siting a high-level nuclear waste repository for the United States. After criticism of the Department of Energy's original decision, a thorough study was conducted prior to the Department of Energy selecting the Yucca Mountain, Nevada site. A desirable feature of society is that everyone has the opportunity to voice their opinion. Even with very thorough study, site selection problems are extremely difficult, as opposition usually arises to assure that such facilities are placed "not in my back yard" late in the analysis. That has been the outcome of the Yucca Mountain repository, as the voters of Nevada have clearly gone on record in opposition to the use of the facility at that location (Ginsburg, 1995). This case demonstrates the need to thoroughly consider risk aspects of nuclear disposal sites.

A decision related to siting the high-level nuclear waste repository was generated by the strategic arms limitation treaties (Dyer, et al., 1998). In the early 1990s the U.S. Department of Energy needed to decide on the process to be used to dispose of excess plutonium. The decision analysis process using multiattribute utility theory is described in Chapter 4. This decision has been made, and initial implementation has just begun. Both this and the high-level waste repository decisions reflect the dimension of time involved in waste disposal. The standard of analysis adopted by the U.S. Government is 10,000 years. This is due for the most part because of the very long half-life of plutonium. The immense degree of uncertainty created by this time frame leads to extraordinary levels of uncertainty in almost all aspects of analysis.

Among the many implementations of AHP are some facility location decisions. AHP has been used to support Finnish study of power generation alternatives (Hämäläinen, 1988; 1990; 1991), location of pipelines (Dey, et al., 1994; 1996; 1998), and water resource planning (Srinivasa Raju and Pillai, 1999a; 1999b). The last two applications are reviewed in Chapter 5.

Chapter 6 involves analysis of more mundane forms of waste in Finland. There have been a number of articles presenting the application of outranking methods to this problem, supporting local government decision-making. The operation of the ELECTRE outranking method is demonstrated in Chapter 6. Other studies of the same problem have used multiattribute utility theory (Pavlov, et al., 1996).

The last in-depth application of multiple criteria analysis to site location decisions reports pipeline location for the transportation of gas and petroleum (Chapter 7). The examples analyzed in Chapter 7 were in Russia and the United States, using a qualitative methodology, Verbal Decision Analysis (Larichev and Moshkovich, 1997). Ordinal scales of criteria with estimates expressed in language natural to decision makers and groups active in the decision are considered in all stages of the method. The manner in which this approach supports group decision-making is emphasized in that categorical measures of criterion accomplishment are considered using a method developed by one of this book's authors. The support of this approach to group decision-making is emphasized.

Chapter 8 reviews the methods demonstrated, with comparative analysis of how they deal with various aspects of multiple criteria. The problems of supporting uncertainty and group decision environments are analyzed as well. Each method has relative strengths. We discuss how different tools would be expected to perform in different problem environments. The nature of the problem to a large extent defines which tool would be best.

Chapter 9 presents overall conclusions.

References

Abu-Taleb, M.H. & Mareschal, B. Water resources planning in the Middle East: Application of the PROMETHEE V multicriteria method. *European Journal of Operational Research*, vol. 81, no. 3 1995, 500-511

Al-Kloub, B., Al-Shemmeri, T. & Pearman, A. The Role of Weights in Multi-Criteria Decision Aid, and the Ranking of Water Projects in Jordan. *European Journal of Operational Research*, vol. 99 1997, 278-288.

Al-Shemmeri, T., Al-Kloub, B. & Pearman, A. Computer Aided Decision Support System for Water Strategic Planning in Jordan. *European Journal of Operational Research*, vol. 102, 1997, 455-472.

Anandalingam, G. A Multiple Criteria Decision Analytic Approach for Evaluating Acid Rain Policy Choices. *European Journal of Operational Research*, vol. 29, 1987, 336-352.

Azis, I.J. Analytic hierarchy process in the benefit-cost framework: a post-evaluation of the trans-sumatra highway project. *European Journal of Operational Research*, vol. 48, no. 1, 1990, 38-48.

Barbarosoglu, G. and Pinhas, D. Capital rationing in the public sector using the analytic hierarchy process. *Engineering Economist*, vol. 40, no. 4, 1995, 315-341.

Barda, O.H., Dupuis, J. & Lencioni, P. Multicriteria Location of Thermal Power Plant. *European Journal of Operational Research*, vol. 45, 1990, 332-346.

Briggs, T., Kunsch, P.L. & Mareschal, B. Nuclear Waste Management: An Application of the Multicriteria PROMETHEE Methods. *European Journal of Operational Research*, vol. 44, no. 1, 1990, 1-10

Brown, C.A. The central Arizona water control study: A case for multiobjective planning and public involvement. *Water Resources Bulletin*, vol. 20, no. 3, 1984, 331-337.

Cohon, J. *Multiobjective Programming and Planning*, Academic Press, New York, 1978.

Cohon, J., ReVelle, C., Current, J., Eagles, T., Eberhart, R. & Church, R. Application of a multiobjective facility location model to power plant siting in a six-state region of the U.S. *Computers & Operations Research*, vol. 7, 1980, 107-123.

Current, J., Minh, H. & Schilling, D. Multiobjective analysis of facility location decisions. *European Journal of Operational Research*, vol. 49, 1990, 295-307.

Delhaye, C., Teghem, J., and Kunsch, P. Application of the ORESTE method to a nuclear waste management problem. *International Journal of Production Economics*, vol. 24, nos. 1,2, 1991, 29-39.

De Montgolfier J., Bertier P. *Approach multicritere des problemes de decision*, Editions Hommes et Techniques, Paris (in French), 1978.

Dey, P., Tabucanon, M.T., and Ogunlana, S.O. Planning for project control through risk analysis: A Petroleum pipeline-laying project. *International Journal of Project Management*, vol. 12, no. 1, 1994, 23-33.

Dey, P.K.. Tabucanon, M.T., and Ogunlana, S.O. Petroleum pipeline construction planning: A conceptual framework. *International Journal of Project Management*, vol. 14, no. 4, 1996, 231-240.

Dey, P.K. Ogunlana, S.O. Gupta, S.S., and Tabucanon, M.T. A risk-based maintenance model for cross-country pipelines. *Cost Engineering*, vol. 40, no. 4, 1998, 24-31.

Dyer, J.S., Edmunds, T., Butler, J.C., and Jia, J. A multiattribute utility analysis of alternatives for the disposition of surplus weapons-grade plutonium. *Operations Research*, vol. 46, no. 6, 1998, 749-762.

Easterling, D. and Kunreuther, H. *The Dilemma of Siting a High-Level Nuclear Waste Repository*, Kluwer Academic Publishers, Boston, 1995.

Flanders, N.E., Brown, R.V., Andre-eva, Y. and Larichev, O. Justifying public decisions in Arctic oil and gas development: American and Russian approac· ·s. *Arctic*, vol. 51, no. 3, 1998, 262-279.

Georgopoulou, E., Sarafidis, Y., and Diakoulaki, D. Design and implementation of a group DSS for sustaining renewable energies exploitation. *European Journal of Operational Research*, vol. 109. 1998, 483-500.

Ginsburg, S. *Nuclear Waste Disposal: Gambling on Yucca Mountain*, Aegean Park Press, 1994.

Grassin, N. Constructing 'population' criteria for the comparison of different options for a high voltage line route. *European Journal of Operational Research* 26, 1986, 42-57.

Haastrup, P., Maniezzo, V., Mattarelli, M., Mazzeo Rinaldi, F., Mendes, I., and Paruccini, M. A decision support system for urban waste management. *European Journal of Operational Research*, vol. 109, 1998, 330-341.

Hämäläinen, R.P. Computer assisted energy policy analysis in the Parliament of Finland, *Interfaces* 18:4, 1988, 12-23.

Hämäläinen, R.P. A decision aid in the public debate on nuclear power, *European Journal of Operational Research* 48:1, 1990, 66-76.

Hämäläinen, R.P. Facts or values – how do parliamentarians and experts see nuclear power? *Energy Policy* 19:5, 1991, 464-472.

Inhaber, H. "How much for this dump?" *Across the Board*, December 1992, 33-38.

Jones, M., Hope, C., and Hughes, R. a multi-attribute value model for the study of UK energy policy. *Journal of the Operational Research Society*, vol. 41, no. 10, 1990, 919-929.

Karagiannidis, A. & Moussiopoulos, N. Application of ELECTRE III for the Integrated Management of Municipal Solid Wastes in the Greater Athens Area. *European Journal of Operational Research* 97 1997, 439-449.

Keeney, R.L., McDaniels, T.L. and Swoveland, C. Evaluating Improvements in Electric Utility Reliability at British Columbia Hydro. *Operations Research* 43(6) 1995, 933-947.

Keeney, R.L., von Winterfeldt, D. and Eppel, T. Eliciting public values for complex policy decisions. *Management Science*, vol. 36, no. 9, 1990, 1011-1030.

Kim, T.-Y., Kwak, S.-J., and Yoo, S.-H. Applying multi-attribute utility theory to decision making in environmental planning: A case study of the electric utility in Korea. *Journal of Environmental Planning & Management*, vol. 41, no. 5, 1998, 597-609.

Kleindorfer, P.R. and Kunreuther, H.C. *Insuring and Managing Hazardous Risks: From Seveso to Bhopal and Beyond*, Springer-Verlag, Berlin, 1987.

Kunreuther, H. and Easterling, D. Are risk-benefit tradeoffs possible in siting hazardous facilities? *The American Economic Review*, vol. 80, no. 2, Papers and Proceedings of the Hundred and Second Annual Meeting of the American Economic Association (May, 1990), 1990, 252-256.

Kunreuther, H.C., Linnerooth, J., Lathrop, J., Atz, H., Macgill, S., Mandl, C., Schwarz, M. and Thompson, M. *Risk Analysis and Decision Processes: The Siting of Liquefied Energy Gas Facilities in Four Countries*. Springer-Verlag, Berlin, 1983.

Larichev, O. *Objective Models and Subjective Decisions*, Nauka Publishing House, Moscow, 1987 (in Russian)

Larichev, O. and Moshkovich, H. *Verbal Decision Analysis for Unstructured Problems*, Kluwer Academic Publishers, Boston, 1997.

Medvedev, Z. *The Legacy of Chernobyl*, W. W. Norton & Company, New York, 1990.

Mladineo, N., Margeta, J., Brans, J.P. & Mareschal, B. Multicriteria Ranking of Alternative Locations for Small Scale Hydro Plants. *European Journal of Operational Research* 31 1987, 215-222.

Mladineo, N., Lozic, I., Stosic, S., Mlinaric, D., and Radica, T. An Evaluation of Multicriteria Analysis for DSS in Public Policy Decision. *European Journal of Operational Research* 61(1,2) 1992, 219-229.

Pavlov, Y., Grancharov, D. Momchev, V. Economic and ecological utility oriented analysis of the process of anaerobic digestion of waste waters. *European Journal of Operational Research*, vol. 88, no. 2, 1996, 251-256.

Powell, J.C. The evaluation of waste management options. *Waste Management & Research*, vol. 14, 1996, 515-526.

Renn, O., Stegelmann, H.U., Albrecht, G., Kotte, U. and Peters, H.P. An empirical investigation of citizens' preferences among four energy scenarios. *Technological Forecasting and Social Change* 26, 1984, 11-46.

Roy, B. and Bouyssou, D. Comparison of two decision-aid models applied to a nuclear power plant siting example. *European Journal of Operational Research* 25, 1986, 200-215.

Salminen, P., Hokkanen, J. and Lahdelma, R. Comparing multicriteria methods in the context of environmental problems. *European Journal of Operational Research* vol. 104, no. 3, 1998, 485-496.

Smith, V.K. and Desvousges, W.H. The value of avoiding a LULU: Hazardous waste disposal sites. *The Review of Economics and Statistics*, vol. 68 no. 2, 1986, 293-299.

Solomon, B. & Haynes, K. A survey and critique of multiobjective power plant siting decision rules. *Socio-Economic Planning Sciences*, vol. 18, no. 2, 1984, 71-79.

Srinivasa Raju, K.S., and Pillai, C.R.S. Multicriterion decision making in river basin planning and development. *European Journal of Operational Research*, vol. 112, no. 2, 1999a, 249-257.

Srinivasa Raju, K.S., and Pillai, C.R.S. Multicriterion decision making of an irrigation system. *European Journal of Operational Research*, vol. 112, no. 2, 1999b, 479-488.

Toland, R.J., Kloeber, J.M., Jr., and Jackson, J.A. A comparative analysis of hazardous waste remediation alternatives. *Interfaces*, vol. 28, no. 5, 1998, 70-85.

Vuk, D., Kozelj, B. & Mladineo, N. Application of multicriterional analysis on the selection of the location for disposal of communal waste, *European Journal of Operational Research*, vol. 55, 1991, 211-217.

Walker, W.E., Abrahamse, A., Bolten, J., Kahan, J.P., van de Riet, O, Kok, M., and den Braber, M. A policy analysis of Dutch river dike improvements: Trading off safety, cost, and environmental impacts. *Operations Research*, vol. 42, no. 5, 1994, 823-836.

2 METHODS AND DECISION PROCESSES: DESCRIPTIVE AND NORMATIVE

The Descriptive Studies

How are large-scale siting decisions made in practice? H. Kunreuther, J. Linneroth, and some of their colleagues (1983) investigated large-scale siting decisions while working in the International Institute for Applied System Analysis (IIASA). The problem under study was the decision process used to site liquified energy gas (LEG) facilities in four countries: the Federal Republic of Germany, the Netherlands, the United Kingdom, and the United States. Different countries have different institutional settings for deciding how siting choices are made. Despite major differences in approaches, Kunreuther and Linnerooth developed a general framework for describing siting decisions.

Kunreuther, et al. (1983) found that different active parties usually participate in siting decisions. Each active party may focus on different criteria (for example, environmental aspects, economic factors, or safety considerations) with different information upon which each party bases their preferences. Some active parties would benefit a great deal from adoption of a project, while others might be harmed. For example, some residents might benefit from nearby location of a LEG terminal. However, other people might have to be relocated to a safer location.

Four primary participant groups in the choice process can be identified:

Government agencies – whose regulatory actions influence public preferences and provide advantages to some active groups. Government agencies are controlled by legislation.

Applicants – firms who support construction and operation of a facility. Applicants base their decision making primarily on economic evaluation, considering project costs and benefits to their operations.

Local residents – people living near the location of a proposed LEG terminal usually have strong feelings about projects of this nature. Facilities can provide jobs, but they also can impose risks of accident or environmental pollution.

Public interest groups – a number of active parties represent the interests and preferences of different components of the overall public.

It is clear that there is potential for conflict between active groups. The degree of conflict depends on a number of factors, such as the distribution of power and responsibility between groups, memory of past disasters associated with a particular technology, the degree of influence on the public of information distributed by various active groups, and so on.

Kunreuther and his colleagues proposed a descriptive framework presenting a typical choice process in siting decisions. They named this framework the Multi-Attribute Multi-Party (MAMP) model. The presentation of public choice as an interaction of many active parties (or groups) each with their own objectives is typical for political science (March and Simon, 1958). The decision process itself is sequential and new issues appear after the partial resolution of previous issues, changing the positions of active parties. Observations of decision processes of this type indicate that a number of rounds typically occur. Each round begins when either:

a) some decision is made

b) a stalemate is reached due to conflict between active parties

c) a new event changes the situation.

At the beginning of each round, some limited set of alternatives is presented. Each active party evaluates some alternatives in terms of the criteria important to them, and on the basis of the information that they have. Subsequently, there is a stage of interaction between parties. At the end of the round, a decision emerges, or decision is stalled due to clear disagreements between active parties.

To represent this decision process, Kunreuther and Linnerooth proposed two useful tools: 1) a matrix of actors and criteria, and 2) a diagram similar to those used in project scheduling presenting rounds and sequential relationships. As an example, for the Nuclear Waste Siting decision we will cover in Chapter 3:

Criteria	**Parties – High Level Nuclear Waste Siting Decision**			
	Dept. of Energy	National Academy of Sciences	National Labs	State of Nevada
Public Health & Safety	x	x		x
Environment	x	x		x
Socioeconomics				x
Cost	x		x	
Technical	x	x	x	

Primary decision responsibility was vested in the Department of Energy. National laboratories provided technical expertise to the Department of Energy. The original decision was made following normal governmental administrative procedures. The National Academy of Sciences (NAS) represents science, in this case seeking sound analysis based on evidence as accurate as could be obtained. The NAS felt very strongly that the Department of Energy's original decision was not based on sound analysis. These concerns led to a new study. The State of Nevada (and subordinate government units) represented local citizens. This group was relatively passive in the early stages of the decision. Many citizens in the immediate vicinity of Yucca Mountain were (and continue to be) quite supportive of siting the facility at that location. However, once the decision was published, it was soon clear that the majority of citizens in Nevada opposed the decision. The State of Nevada has been very active since that time in opposing the Yucca Mountain site.

Another schematic that can aid in understanding a difficult siting decision is a network diagram of the decision process. Kunreuther, et al. (1983) presented such networks in the form of diagrams showing predecessor and concurrent activities. Siting locations, because of their disputative nature, usually involve a series of rounds, or decision processes that result in temporary decisions that are opposed by those who are adversely affected by the specific adopted decision.

PRELIMINARY:

P1 1963   Atomic Energy Commission developed Salt Vault projects in Lyons, KS

P2 1973   Salt Vault project dropped due to technical problems and local opposition

P3 1973   Carlsbad, NM officials invited the Atomic Energy Commission to study salt formations

P4 1973   Energy Research and Development Administration takes over responsibility for disposal

P5 1976   ERDA initiates studies of promising geologic formations in 36 states, intending regional disposal

PHASE I:

A1 1977   President Carter adopts policy of not relying on reprocessing of spent nuclear fuel; DOE task force formed to develop an alternative policy

A2 Feb 1978   DOE task force endorsed geologic disposal in principal; Interagency Review Group on Nuclear Waste Management formed, consisting of top DOE officials and officials of other governmental agencies

A3 1979   Interagency Review Group recommended geologic disposal

A4 1980   DOE study published recommending geologic repository over other options considered

A5 1982   Nuclear Waste Policy Act officially designated geologic disposal for permanent disposal of high-level nuclear waste

B   National Academy of Sciences reviews process, finds it
                     biased and unsatisfactory

PHASE II:
Eastern US
C1 1983-1985 DOE surveys crystalline rock formations in the upper
             Midwest, New England, and the south Atlantic states
C2 Jan 1986 DOE announces 12 potentially acceptable sites in
             Minnesota, Wisconsin, Maine, New Hampshire, Virginia,
             North Carolina, and Georgia
C3 May 1986 DOE suspends search for eastern repository
C4 Apr 1987 DOE announces that the search for an eastern repository
             will resume, but not until 2007

Western US
D1 Spring 1983 Nine previously studied sites are announced as being
               considered
D2 Dec 1984 Yucca Mountain, Hanford, Deaf Smith County, Davis
             Canyon, and Richton Dome nominated for site
             characterization by DOE
D3 May 1986 Yucca Mountain, Hanford, and Deaf Smith County selected
             for site characterization

E1 Dec 1987 Congress names Yucca Mountain as the only site to be
             considered
E2 1989   State of Nevada tries to eliminate Yucca Mountain from
             consideration by passing anti-repository laws and filing
             suit against DOE
E3 1990-1991 Federal courts disallow the actions of the State of Nevada
             against using Yucca Mountain
E4 1984-1992 EPA queried about the licensability of Yucca Mountain

F1 1992   Energy Policy Act eases standards for licensing Yucca
             Mountain
F2 1992-1994 Site characterization delays make the 2010 opening date
             questionable
F3 1992-1994 Calls for review of the repository program

G   1994   DOE proposes accelerated analysis of Yucca Mountain, and requests the Nuclear Regulatory Commission to grant a tentative license with less stringent standards

H   Opposition from State of Nevada precludes use of facility

A network shows the predecessor relations of the decision process. In this case, it traces an historic sequence of events.

Sequence of Major Events: High Level Nuclear Waste Repository

PHASE I

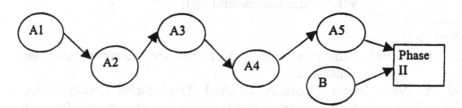

PHASE II
Eastern US

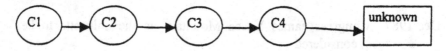

Western US

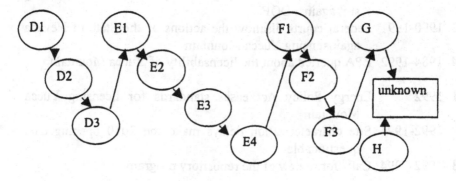

This network displays only some of the publicized activities involved in the high-level nuclear waste repository siting decision. Most of the network is a sequence of governmental activities. Exceptions are activities B (National Academy of Science study) and H (response of the citizens of Nevada). These intervening activities were certainly not part of the planned sequence of events, and had major roles in redirecting the siting process.

The MAMP model can be viewed as a model of conflict resolution between the active parties. Each active party in the discussion process demonstrated a typical decision making behavior – bounded rationality (Simon, 1960). Decision-makers reach conclusions based upon the information they have about the impact of project proposals, and their understanding of the impact of projects on their interests.

Viewing solution of multicriteria location problems from the behavioral perspective, we could emphasize that there is a combination of individual and collective choice. Representatives of each active group are decision-makers, with the usual restrictions of rationality typical for individual decision-making processes (Kahneman, et al., 1982). In other words, these active group participants demonstrate the use of human heuristics and biases during decision-making.

These descriptive studies lead us to the following conclusions:

1. Conflict between active groups is a typical feature of siting problems.
2. The decision process can be presented in a MAMP framework.
3. The behavior of each decision-maker is defined by the limitations and potential of the human information processing system.

Now let us present the normative tools prescribing how people should make siting decisions.

Sample Problem

We generate a sample problem to demonstrate the most popular of the multicriteria tools available for analysis of site location decisions. The sample problem would involve a number of decision participants, from government, local citizenry, and the power industry. The Department of Energy (DOE) would be expected to have primary decision making authority. They would be responsible to represent the interests of the U.S. government and all of its citizens. The Environmental Protection Agency (EPA) would also represent the interests of citizens, primarily in the area of ecological concerns. National Laboratories are subordinate to the Department of Energy, and while they do not have a direct interest, they are the primary source of expertise. Laboratories also have a vested interest in developing useful work for the future. Local citizens are de facto part of the decision making. While they were not active during the initial decision making in the Yucca Mountain case, they have forced the ultimate decision to date by the refusal of the State of Nevada to allow the Yucca Mountain repository to operate. Finally, the Power Industry is a decision making participant. It is assumed that their role is to maximize return to stockholders, and thus to focus on cost aspects of the decision. However, wise commercial decision-making today clearly needs to consider other factors, such as the environment. An example party/concern matrix for the sample problem could be:

Party/Concern Matrix: Sample Problem				
Concerns	DOE	EPA	Local Citizens	Power Industry
Expected deaths	x		x	
Aesthetics		x	x	
Biological impact		x	x	
Socioeconomic	x		x	
Construction cost	x			x
Operating cost				x
Risk	x	x	x	x

In the hypothesized example, the sequence of events could have included the following, based on Y0 as the year in which the decision problem was first considered.

#	Date	Actor	Event
1	Y0	Power Industry	The power industry files suit with the government to provide a permanent

repository to store the nuclear waste that has been accumulating.

2 Y2   Congress   An act funding a study is passed through Congress, giving the Department of Energy the responsibility of conducting the study.

3 Y3   DOE   The DOE commissions a study of the scope of the problem, to identify the quantity of waste involved, its toxicity, and desired features of a permanent storage site.

4 Y3   National Labs   Conduct the DOE-commissioned study, identifying quantities, toxicities, and desired permanent storage site features.

5 Y4   DOE   Commissions a study to recommend an initial set of sites for consideration.

6 Y4   EPA   Conducts environmental impact analyses of the sites proposed.

7 Y5   DOE   Organizes hearings including power industry representatives, as well as members of the public from both proposed sites and from power plant locations.

8 Y6   DOE   Conducts site analysis, calling on National Labs for technical support.

9 Y7   DOE   Announces that three sites (in Tennessee, New Mexico, and California) are being considered for the proposed site.

10 Y7 DOE   Holds hearings in Tennessee, in New Mexico, and in California in order to gauge the feelings of the public in these three states.

11 Y8 DOE   Announces the selection of the Tennessee site.

12 Y9 Congress   Considers bills proposed by Senators and Congresspersons from Tennessee concerning public safeguards related to the proposed repository.

13 Y11 DOE   Obtains budget approval to begin the project construction at the Tennessee site.

Thereafter   Unknown

The network chart for this process could be:

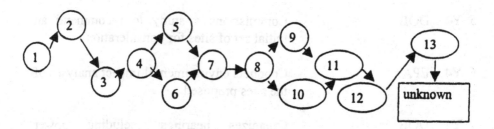

Resolution of negotiation and ultimate decision making must rely upon an evaluative process. It should not be based entirely on the conclusions of any decision analysis. However, these studies (supplemented by party/concern matrices and sketches showing the sequence of events) can help clarify the decision process and lead to more complete consideration of important decision factors. Note that the decision ends in an unknown state. This reflects only part of the large level of uncertainty. Interventions by other parties can have a major impact in changing the siting decision process.

Siting Project Analysis

We will use this general scenario as the basis for a demonstration problem with which to show how the multiple criteria analytic methods that are presented in the subsequent chapters work. We will also demonstrate cost-benefit analysis, the primary means of analysis used by the Corps of Engineers and some other U.S. governmental agencies.

In the sample problem three sites are under consideration for disposal of noxious wastes. One is located in Tennessee, the second in New Mexico, and the third in California. Criteria to be considered are expected number of deaths from the system, aesthetic impact, impact on flora and fauna, socioeconomic impact, construction costs, operation costs, and risk. Measures for each option are given in Table 1.

Table 1: Measures of Criteria Performance by Alternative

Criteria	Tennessee	New Mexico	California
Expected deaths	50	30	180
Aesthetic	$10 million	$4 million	$18 million
Biological	12 species	40 species	30 species
Socioeconomic	20,000 people	10,000 people	50,000 people
Construction costs	$20 billion	$18 billion	$30 billion
Operating costs	$15 billion	$18 billion	$10 billion
Risk	700 years	300 years	80 years

Deaths expected include those incurred due to construction, transportation of material to the site, and exposure of inhabitants in the vicinity of the site. They are estimated over a 100-year period, the expected life of the repository. Aesthetic measures are expressed in terms of the estimated money needed to restore local sites to their current attractiveness to tourists and hikers. Biological data refers to the number of species potentially adversely impacted. Socioeconomic measures estimate the number of jobs impacted by selecting each location. Construction costs and operation costs are estimated in net present terms over the life of the project. Risk is expressed as the expected number of years before a catastrophe might expose materials stored in the repository.

Cost-Benefit Analysis

The basic idea of cost-benefit analysis is to price all elements of the analysis. In principle, this is an ideal method of analysis, as everything is ideally represented by its economic worth. This approach is widely used. One example applied to siting locally undesirable land uses was by Smith and Desvousges (1986). That study provided an example of the concrete output obtainable from cost-benefit analysis. Average subject preferences for each mile between a residence and a landfill containing hazardous waste were calculated at between $330 and $495 annually per household for one specific siting decision. This approach offers an approach to resolve difficult conflicts through compensation of those stakeholders who are adversely affected by a decision. Sarin (1983) presented a study of cost-benefit analysis in a risk decision where impacted constituents were provided incentives for their cooperation. We will see in Chapter 3, however, that the citizens of Nevada take the position that fundamental principles cannot be compensated. There are different views as to the rationality of such a position. Walker, et al. (1994) presented their cost-benefit analysis of Dutch river dike improvements, considering criteria of financial costs and flood damage protection. In practice, of course, many things are difficult to estimate. In the Walker, et al. study, landscape, natural and cultural values were not included in the cost-benefit calculations, but were presented in terms of the expected percentage of features preserved.

Another difficulty with cost-benefit analysis is the treatment of the timing of cash flows. The net present cost for construction is known to involve high levels of error. Regardless, this factor is by far the easiest component to estimate of those criteria included in our demonstration model. The hardest might be aesthetic features.

Expected deaths need to be converted to a dollar value. We arbitrarily take a value for a life of $5 million. These are distributed evenly over the 100-year life of the project, and costs discounted back to net present terms. Estimates are given for aesthetic, biological, and socioeconomic factors as well. We save tremendous amounts of difficulty by presenting the final numbers. Most cost-benefit analyses are criticized by project opponents. The easiest targets to criticize are estimates of intangible items such as these. Nonetheless, our estimates in net present terms are given in Table 2.

Table 2: Net Present Costs for Demonstration Project

Criteria	Tennessee	New Mexico	California
Expected deaths	$22 million	$13 million	$78 million
Aesthetics	$10 million	$4 million	$18 million
Biological	$9 million	$30 million	$23 million
Socioeconomic	$4000 million	$2000 million	$10000 million
Construction costs	$20000 million	$18000 million	$30000 million
Operating costs	$15000 million	$18000 million	$10000 million
Risk	$320 million	$1400 million	$2800 million

Once this stage of the analysis is reached, the cost-benefit approach is very clear and straightforward. If the numbers in Table 2 are accepted, the expected net present costs for the three options are easily identified as the sum of the seven criteria estimates:

Tennessee	$39,361 million
New Mexico	$39,447 million
California	$52,919 million

On the basis of these numbers, the Tennessee site would be selected. The New Mexico site is a very close second. The California site seems far inferior to the other two.

Multiattribute Utility Theory

Cost-benefit analysis, as we have seen, tries to measure everything in terms of economic impact. Economic theory is based on the concept that everything has a price. But not everyone accepts the idea that a dollar spent on one thing is necessarily the same as a dollar spent on something else. Deaths, for instance, may be difficult to compare to money spent on construction. There can be nonlinearities, in that 100 expected deaths might be rated as less than twice as bad as 50 deaths. Combining the seven criteria as done in cost-benefit analysis implies linearity. Aggregating dollars across all criteria by addition also infers that all seven types of dollars are the same in value.

Multiattribute utility (Keeney and Raiffa, 1976) combines all criteria, but in terms of utility (or value) rather than in monetary terms. The utility of possible values on each criterion is determined without consideration of alternatives. Nonlinear utility functions are allowed, providing for cases where, for instance, 100 deaths are worse than 50 deaths, but not twice as bad. An aggregation formula in terms of relative weights is established. If preference independence does not exist, nonlinear terms can be included to reflect contingent relative value.

Single-Attribute Utility Functions
In the demonstration model, utility functions for each criterion can be established on normalized scales where the best possible measure is assigned a value of 1.0 and the worst possible measure is assigned a value of 0. This avoids the need to convert values into their monetary equivalents.

Expected Deaths Utility

best imaginable	0 deaths	utility = 1.0
	30 deaths	utility = 0.8
	50 deaths	utility = 0.5
	180 deaths	utility = 0.3
worst imaginable	1000 deaths	utility = 0

Aesthetic Utilities

best imaginable	no damage	utility = 1.0
	$4 million	utility = 0.92
	$10 million	utility = 0.8
	$18 million	utility = 0.64
worst imaginable	$50 million	utility = 0

Biological Damage Utilities

best imaginable	no damage	utility = 1.0
	12 species	utility = 0.8
	30 species	utility = 0.5
	40 species	utility = 0.4
worst imaginable	200 species	utility = 0

Socioeconomic Utilities

best imaginable	0 affected	utility = 1.0
	10000 people	utility = 0.6
	20000 people	utility = 0.4
	50000 people	utility = 0.3
worst	200000 people	utility = 0

Construction Cost Utilities best imaginable $1 billion   utility = 1
                                                   $18 billion   utility = 0.81
                                                   $20 billion   utility = 0.79
                                                   $30 billion   utility = 0.7
                                  worst imaginable $200 billion utility = 0

Operating Cost Utilities   best imaginable $1 billion   utility = 1
                                                   $10 billion   utility = 0.5
                                                   $15 billion   utility = 0.3
                                                   $18 billion   utility = 0.2
                                  worst imaginable $25 billion   utility = 0

Risk Utilities   best imaginable   2000 years   utility = 1.0
                                                   700 years   utility = 0.9
                                                   300 years   utility = 0.8
                                                   80 years   utility = 0.2
                                  worst imaginable 1 year   utility = 0

Weights for Criteria

The utilities for each criterion can be combined into an overall value score
for each alternative that can be used as a value function to compare
alternatives. Identification of tradeoffs is accomplished in a number of
different ways. Multiattribute utility theory usually applies lottery
tradeoffs, with one form involving the decision-maker comparing one
alternative with the best of all features except for one criterion, and the
second alternative doing the same for some other alternative. Then, the
alternative not selected is improved by the decision-maker until the two
alternatives are equal in value. For instance:

	Alternative 1	Alternative 2
Expected Deaths	**1000**	**0**
Aesthetics damage	**no damage**	**$50 million**
Biological	no damage	no damage
Socioeconomic	0 people affected	0 people affected
Construction costs	$1 billion	$1 billion
Operating costs	$1 billion	$1 billion
Risk of catastrophe	2000 years	2000 years

If the decision-maker selects one alternative, then the decision-maker is asked to improve the other alternative until the two alternatives are equal in value. For instance, assume the decision-maker selected the second alternative, with best performance on expected deaths and worst performance on aesthetics. Then, the decision-maker would be asked how many deaths in the first alternative would make the two alternatives equal in value. A possible response would be that a combination of 30 expected deaths and no aesthetic damage would be equivalent in value to 0 expected deaths and $50 million damage to aesthetics.

	Alternative 1	Alternative 2
Expected Deaths	**30**	0
Aesthetics damage	no damage	$50 million
Biological	no damage	no damage
Socioeconomic	0 people affected	0 people affected
Construction costs	$1 billion	$1 billion
Operating costs	$1 billion	$1 billion
Risk of catastrophe	2000 years	2000 years

The scales of measure on each criterion are considered. A minimum of six sets of pair-wise comparisons are required to develop the full set of weights over the seven alternatives. Checking preference independence would require more pair-wise comparisons. Using the minimum number of pair-wise comparisons would not allow any inconsistency, but in order to be accurate, absolute precision in tradeoff responses is required. Accuracy can be improved by asking additional pair-wise comparisons, although it would also create the need to reconcile probable conflicts. The six pair-wise comparisons of equivalent preference in the sample were:

Expected Deaths	30	0
Aesthetics	no damage	$50 million damage
Expected Deaths	50	0
Biological Damage	0 species	200 species
Expected Deaths	20	0
Socioeconomic	0 people affected	200000 people affected

Expected Deaths	50	0
Construction Cost	$1 billion	$200 billion

Expected Deaths	5	0
Operating Cost	$1 billion	$25 billion

Expected Deaths	20	0
Risk	2000 years	1 year

These tradeoffs are converted to relative weights, reflecting both scale and decision-maker preference. In order to identify the weights, the decision maker must be asked to identify the probability value p that would make the following choices A and B equivalent:

CHOICE A: at probability 1.0
The most important criterion at its best level: **Expected Deaths 0**
All other criteria at their worst levels: **Aesthetics $50 mill dam**
   Biological damage   200 species
   Socioeconomic   200000 people affected
   Construction cost $200 billion
   Operating cost $25 billion
   Risk   1 year

CHOICE B:	at probability p	at probability (1-p)
Expected Deaths	0	**1000**
Aesthetics	**no damage**	**$50 million damage**
Biological damage	0 species	200 species
Socioeconomic	0 people affected	200000 people affected
Construction cost	$1 billion	$200 billion
Operating cost	$1 billion	$25 billion
Risk	2000 years until disaster	1 year until disaster

The most important criterion would be assigned a weight of p, and all other weights calculated relative to that weight. According to MAUT theory (Keeney and Raiffa, 1976; Keeney, 1980) the sum of all criteria weights is to be calculated. When this sum is equal to 1, the utility function is additive; when this sum is not equal to 1 the utility function is multiplicative.

If we assume that the aggregation formula is additive from very beginning of the analysis, it is possible to avoid the last question to the decision maker (see above) and use the condition that the sum of criteria weights is equal to one. Such a supposition is often used in analyses involving government agencies or groups of decision makers.

A set of weights obtained through LOGICAL DECISION for the tradeoffs presented above is:

Value =	0.851 × Expected Deaths	in people
+	0.026 × Aesthetics	in $ million
+	0.043 × Biological damage	in species
+	0.017 × Socioeconomic disruption	in people
+	0.043 × Construction costs	in $ billion
+	0.004 × Operating costs	in $ billion
+	0.017 × risk	in years before catastrophe

Aggregation

These tradeoffs convert the measures of the alternatives into value scores, reflecting both single-attribute utility functions for each measure, as well as a multiattribute utility function reflecting the relative importance of the seven measures (considering their ranges). This yields a ranking of alternatives by value as follows:

Tennessee	value score	0.538
New Mexico	value score	0.781
California	value score	0.334

This differs from the order obtained from cost-benefit calculations, with the New Mexico site now notably more attractive than the Tennessee site. This is clearly because of the New Mexico site's advantage in expected deaths. Tennessee had a cost advantage over New Mexico, but operating costs were discounted with this preference function.

Analytic Hierarchy Process

The Analytic Hierarchy Process (AHP - Saaty, 1977, 1980, 1982) is a technique for converting subjective assessments of relative importance into a set of weights. It has proven to be very useful in assisting selection from a finite set of alternatives using subjective measures (see Zahedi, 1986 and Shim, 1989 for surveys).

AHP is a means of developing measures in physical or social environments when physical or statistical measures are unavailable. In the social environment, AHP is a means of converting subjective assessments into relative values. Three principles are applied in AHP: problems are decomposed by identifying those factors that are important; comparative judgments are made on the decomposed elements of the problem (measures of relative importance are obtained through pair-wise comparison matrices); and these measures are finally recombined into an overall rating of available choices.

AHP has also been used to develop linear utility functions reflecting relative importance of decision objectives in mathematical programming, as well as for ranking things (such as seeding tennis tournaments or selecting the best cities to live in).

Hierarchy Development

The first step of AHP is to develop the hierarchy of criteria and alternatives, which determines the pair-wise comparison matrices required. In this case, the seven criteria given in Table 1 can be grouped into four general areas: expected deaths, impacts to the environment (including aesthetics, biological impact, and socioeconomic impact), costs (construction and operating), and risk. These four general areas can constitute the second layer of the hierarchy, with subcomponents making a third level. Below each level without subcomponents is a bottom level of alternatives (the Tennessee, New Mexico, and California sites).

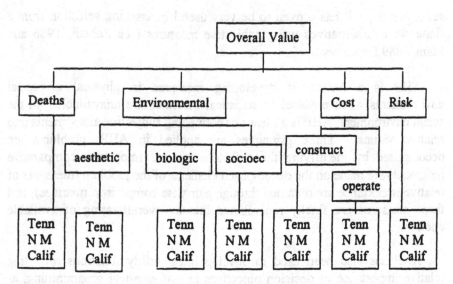

Hierarchies require pair-wise comparison matrices at each node. This is accomplished using a verbal scale. The subjective verbal scale for AHP asks the decision maker to select the preferred item of the pair being considered, and then select the verbal description best describing the decision maker's preference between the two items by this scale:

AHP Verbal Scale for Pairwise Comparisons
1 – Base factor **roughly equivalent** in importance to the other factor
3 – Base factor **moderately** more important than other factor
5 – Base factor **essentially** more important than other factor
7 – Base factor relative importance **very strong**
9 – Base factor **overwhelmingly** more important than other factor

Numbers given below can be viewed as generated using this verbal scale. Here, we need one pair-wise comparison matrix for the four components of Overall Value, one for the three components of environmental criteria, one for the two components of cost criteria, and then seven pair-wise comparison matrices to estimate the relative performance of the three sites on each of the seven measures.

Subjective Pair-wise Comparisons
The next phase in the AHP analysis is to obtain the preference input for the pair-wise comparison matrices. The upper right triangular portion of the

following matrix would be completed with the strength of preference given according to the scale in the adjacent table. (If the row factor is less preferred than the column factor, the reciprocals of these values can be used.) Even (or fractional) numbers can be used for evaluations between the other ratings. Pair-wise comparisons are elicited from the decision-maker, estimating the degree that one factor is preferred over the other. For instance, avoiding deaths might be assessed as essentially more important than environmental concerns. Thus, the pair-wise comparison ratio would be 5, as shown below (along with the other pair-wise comparisons in this matrix):

Base Criterion	Subjective Rating	Compared Criterion	Value
deaths	essentially more important than	environmental	5
deaths	importance very strong over	cost	7
deaths	overwhelmingly more important than	risk	9
environmental	roughly equivalent to moderately	cost	2
environmental	essentially more important than	risk	5
cost	moderately more important than	risk	3

The most commonly used method to obtain relative weights for such a reciprocal matrix is the eigen vector. Another approach would be obtaining the geometric mean, which has theoretical advantages (Barzilai, et al., 1987), as well as being relatively easy computationally. The eigen vector provides a robust estimator, and argument for its justification is provided by Saaty (1977).

Consistency Measure

If each of the ratio pair-wise comparisons given by the decision-maker were precisely accurate, each of the normalized columns would yield identical values. However, when obtaining subjective estimates of relative value, one would expect some variance. That is one reason that multiple estimates of the ratios are obtained (by filling in the entire upper right triangle of pair-wise comparisons). The maximum eigen value provides a measure of inconsistency, which can be used to validate whether the differences in the resulting weights were greater than expected at random. The consistency index (CI) is a function of the maximum eigen value and the number of elements (n) in the pair-wise comparison:

$$CI = (\lambda_{max} - n)/(n-1)$$

where λ_{max} is the maximum eigen value for the pair-wise comparison matrix and n is the number of objectives compared. This inconsistency index can be compared with values expected from a random distribution. Saaty has proposed a cutoff limit of 0.10 of this random index. Software usually adjusts to account for the size of the pair-wise comparison matrix.

Weight Calculations

The assessments provided above yield a pair-wise comparison matrix, with a diagonal identity and reciprocal values below the diagonal:

	Deaths	Environment	Cost	Risk
Deaths	1	5	7	9
Environment	1/5	1	2	5
Cost	1/7	½	1	3
Risk	1/9	1/5	1/3	1

The eigen vector for these pair-wise comparison ratios yields relative weights for the four factors of:

Overall value = 0.664 deaths + 0.184 environment + 0.105 cost + 0.047 risk

The consistency index for these pair-wise comparisons is 0.045, which is less than the limit of 0.1, so the input values appear to be sufficiently consistent. On the hierarchy, environment consists of three subcomponents: aesthetics, biological, and socioeconomic impacts. The relative importance of these three subcomponents together to the overall decision is 0.184. This relative weight must now be allocated over its three subcomponents. The decision-maker might make the assessments:

Aesthetics is rated as ½ times as important as biological impact
Aesthetics is rated as 2 times as important as socioeconomic impact
Biological is rated as 3 times as important as socioeconomic impact

The eigen vector for these assessments yield:

Environment value = 0.297 aesthetics + 0.540 biological + 0.163 socioeconomic

Since environment is 0.184 of the Overall value function, the relative contribution of these three components of environment to Overall value are obtained by multiplying 0.184 times each of the eigen vector values obtained:

> 0.055 aesthetics + 0.099 biological + 0.030 socioeconomic
> consistency index = 0.008

Costs were 0.105 of the Overall value function. Given that construction costs were considered to be 9 times as important as operating costs, an eigen vector for the two components of cost would be:

Cost value = 0.9 construction + 0.1 operating
consistency index = 0.000

The contribution of the two cost components to overall value would therefore be 0.105 times these values, or:

> 0.0945 construction + 0.0105 operating

The Overall value function obtained with AHP is thus:

Overall value =		0.664 expected deaths
	+	0.055 aesthetic impact
	+	0.099 biological impact
	+	0.030 socioeconomic impact
	+	0.0945 construction cost
	+	0.0105 operating cost
	+	0.047 risk

The relative contribution of each alternative to each of these seven criteria now needs to be assessed. This requires seven small 3 by 3 pair-wise comparison matrices comparing how the Tennessee, New Mexico, and California sites perform on each criterion.

Expected deaths:

Tennessee	is	½ times as good as	New Mexico
Tennessee	is	4 times as good as	California
New Mexico	is	½ times as good as	California

ED = 0.323 Tennessee + 0.588 New Mexico + 0.089 California
consistency index = 0.008

Aesthetic impact:

Tennessee	is	1/3 times as good as	New Mexico
Tennessee	is	2 times as good as	California
New Mexico	is	5 times as good as	California

AI = 0.230 Tennessee + 0.648 New Mexico + 0.122California
consistency index = 0.003

Biological impact:

Tennessee	is	4 times as good as	New Mexico
Tennessee	is	3 times as good as	California
New Mexico	is	½ times as good as	California

BI = 0.625 Tennessee + 0.136 New Mexico + 0.238 California
consistency index = 0.016

Socioeconomic:

Tennessee	is	½ times as good as	New Mexico
Tennessee	is	3 times as good as	California
New Mexico	is	5 times as good as	California

S = 0.309 Tennessee + 0.582 New Mexico + 0.109 California
consistency index = 0.003

Construction cost:

Tennessee	is	1 times as good as	New Mexico
Tennessee	is	2 times as good as	California
New Mexico	is	2 times as good as	California

CC = 0.400 Tennessee + 0.400 New Mexico + 0.400 California
consistency index = 0.000

Operating cost:

Tennessee	is	2 times as good as	New Mexico
Tennessee	is	1/4 times as good as	California
New Mexico	is	1/6 times as good as	California

OC = 0.193 Tennessee + 0.106 New Mexico + 0.701 California
consistency index = 0.008

Risk of catastrophe:

Tennessee	is	3 times as good as	New Mexico
Tennessee	is	9 times as good as	California
New Mexico	is	3 times as good as	California

RC = 0.692 Tennessee + 0.231 New Mexico + 0.077 California
consistency index = 0.000

These assessments yield the equivalent of single-attribute utility functions. The AHP input, using integer values from 1 to 9, provides a quick and easy way to express subjective differences quantitatively. This ease comes with some obvious error in precision. For instance, in assessing relative construction costs, Tennessee has a cost of $20 billion, and New Mexico an expected cost of $18 billion, which are quite close, and were assessed as equal (the closest assessment if you are limited to integers and their fractions). But there was a difference of $2 billion. The assessment shows them to be exactly equal, and the consistency index is perfect at 0.000. However, the consistency index merely checks for input consistency, not input accuracy.

Aggregation

The final values for each of the three alternatives are obtained by multiplying the Overall value function derived above times the three by seven matrix of values for each alternative on each criterion. This yields Overall value functions for each alternative of:

Tennessee	0.371
New Mexico	0.507
California	0.123

These values differ from those obtained from MAUT, but are in the same order, and are roughly proportional. New Mexico is again the apparent preferred solution.

Outranking Method

The next method we demonstrate applies the principles of outranking, as developed by the French school at LAMSADE (Roy, 1968). The version we demonstrate here is ELECTRE III.

ELECTRE III

ELECTRE III begins with the data, as shown in Table 1. This data is all quantitative. The method calculates outranking relationships on the basis of concordance indexes reflecting relative advantage of one alternative over another, and discordance reflecting disadvantage of the base alternative to the other alternative. For instance, the alternatives Tennessee and New Mexico can be compared. The concordance index for Tennessee would consider those criteria for which Tennessee has an advantage over New Mexico.

Criteria	Advantage
Biological	Tennessee 12 species, New Mexico 40 species
Operating costs	Tennessee $15 billion, New Mexico $18 billion
Risk	Tennessee 700 years, New Mexico 300 years

Conversely, the concordance index for New Mexico would consider those criteria for which New Mexico has relative advantages over Tennessee:

Criteria	Advantage
Expected deaths	New Mexico 30, Tennessee 50
Aesthetics	New Mexico $4 million, Tennessee $10 million
Socioeconomic	New Mexico 10,000 people, Tennessee 20,000
Construction costs	New Mexico $18 billion, Tennessee $20 billion

Outranking calculation depends on weights and threshold parameters. Weights are entered directly by the decision-maker, and are used to calculate the degree of relative advantage of one alternative over the other. Here weights were entered as follows:

Criterion	Weight	Tennessee	New Mexico
Expected deaths	0.80		better
Aesthetics	0.05		better
Biological	0.10	better	
Socioeconomic	0.05		better
Construction cost	0.10		better
Operating cost	0.10	better	
Risk	0.10	better	

Threshold limits are the preference coefficient p, the indifference coefficient q, and the veto coefficient v. Those parameters are needed for each criterion used in the pair-wise comparison of two alternatives (A and B). These parameters are used to determine strong outranking and weak outranking relationships of the base alternative to the alternative it is being compared to. A **strong preference** is identified when the performance on that criterion of alternative A is better than the performance of alternative B by p or more. A **weak preference** for the alternative A on that criterion would be identified when the relative performance of alternative A over alternative B was greater than q but less than p. The alternatives would be **indifferent** in preference on the criterion being evaluated if the difference in performance between the two alternatives is less than q. If the performance of alternative B on one criterion is better than the veto coefficient v, then the performance of alternative A could not outrank alternative B.

The **concordance index** of alternative A relative to alternative B is the proportion of the sum of the weights of those criteria for which the base alternative has a strong preference over the other alternative, plus a portion of the weights of those criteria for which the base alternative has a weak preference over the other alternative, divided by the sum of all criteria weights.

For instance, we can use Tennessee as the first alternative, and compare it to the New Mexico site. Coefficients p, q, and v need to be determined. These can be entered as a constant difference, or as a proportional difference, or as a function including both types of difference. They could be defined by the relation to the best estimate on the criterion (direct estimation) or by the relation to the worst estimate (inverse estimation). The veto coefficient does not have to be defined for all criteria.

Let us suppose that the decision maker gave the following information (coefficients are given as constant differences).

Criteria	Expected deaths	Aesthe tics	Biolog ical	Socioec onomic	Const cost	Op cost	Risk
Weight	**0.8**	**0.05**	**0.1**	**0.05**	**0.1**	**0.1**	**0.1**
Level of indifference q	10	$ 1 million	5 species	1000 people	1 billion	1 bill	20 years
Level of preference p	40	$4 million	15 species	5000 people	4 billion	2 bill	100 years
Veto v	90		25 species	10000 people			

This implies that if the estimate of one alternative is no better or worse than the other alternative on one criterion than the q level of indifference, the decision maker is indifferent to their relative performance. If the difference in performance is greater than the p level of preference, the decision maker has a strong preference for the better performing alternative. The p and q coefficients are used to determine the concordance indices. The p and v coefficients are used to determine discordance indices.

Tennessee is better than New Mexico on biological impact, operating cost, and risk. The concordance index contribution is the sum of the weights of those criteria where the first alternative outperforms the other alternative by at least p. For those criteria where the base alternative outperforms the other alternative by more than q but less than p, a portion of criterion weight is added to the concordance index. This proportion is calculated:

$$((\text{difference in performance}) - q) / (p - q)$$

The concordance coefficient for Tennessee versus New Mexico equals:

$$C(\text{T-NM}) = [0.1 \times 1 + 0.1 \times 1 + 0.1 \times 1] / 1.3 = 0.23$$

Tennessee is worse than New Mexico on expected deaths, aesthetic impact, socioeconomic impact, and construction cost. The **discordance index** reflects the greatest degree of disadvantage of the first alternative to the other alternative. If the disadvantage of the first alternative on a criterion is less than p, there is no contradiction with the statement that the first alternative is better than the other alternative. The discordance index is calculated only for those criteria having a veto parameter. If the disadvantage of the base alternative is greater than v, the discordance index for that criterion is 1.0. If the disadvantage is between p and v, the discordance value is proportional to (disadvantage $- p$) $/ (v - p)$. The discordance index would be the greatest of these discordance values. In comparing the Tennessee site against the New Mexico site:

Criteria	Tennessee	New Mexico	degree	
Expected deaths	50	30	disadvantage less than p	0.0
Socioeconomic	20,000 people	10,000 people	disadvantage equal to v	1.0

The greatest of these values is 1.0, implying that the discordance index of the Tennessee site compared to the New Mexico site is: D(T-NM) = 1.0.

The **credibility degree** reflects the truth of the statement that one alternative is better than the other. If the discordance value over all criteria is less than the concordance index, the credibility degree is equal to the concordance index. The concordance index for the Tennessee site relative to the New Mexico site is 0.23, while two of the discordance values are equal to 1.0. So we need to consider the other case.

When one or more discordance values exceed the concordance index, the credibility degree is equal to:

Concordance index $\times \Pi_k$ [(1 – discordance value) / (1 – concordance index)]

where k is the criterion index.

For Tennessee compared to New Mexico, this calculation would be:

$$0.23 \times [(1 - 1.0) \times (1 - 0.0) / (1 - 0.23)] = 0.$$

Thus the Tennessee site has a credibility value of 0 relative to the New Mexico site.

Let us now compare the New Mexico site with the Tennessee site. Using the same p, q, and v parameters as above, we can calculate the concordance index C(NM-T). New Mexico is better than Tennessee on the criteria expected deaths, aesthetic impact, socioeconomic impact, and construction cost.

Criteria	New Mexico	Tennessee	degree
Expected deaths	30	50	advantage between q and p **Outranking weak**
Aesthetics	$4 million	$10 million	advantage greater than p **Outranking strong**
Socioeconomic	10,000 people	20,000 people	advantage greater than p **Outranking strong**
Construction costs	$18 billion	$20 billion	advantage between q and p **Outranking weak**

The proportional construction cost advantage is calculated:

$$C(NM\text{-}T) =$$
$$[0.8\times((50-30)-10)/(40-10) + 0.05\times1.0 + 0.05\times1.0 + (0.1\times((20-18)-1)/(4-1)]/1.3$$
$$= 0.4$$

The discordance index is calculated in the same manner as before. Tennessee is better than New Mexico on biological impact, operating cost, and risk. The only one of these criteria with a veto level is biological impact.

Criteria	New Mexico	Tennessee	degree
Biological	40 species	12 species	disadvantage greater than v 1.0

Therefore, the discordance index of the New Mexico site compared to the Tennessee site is: D(NM-T) = 1.0. The credibility degree for the New Mexico site compared to the Tennessee site is easily seen to be 0.

This implies that while Tennessee cannot be established to be better than the New Mexico site, neither can the New Mexico site be established as superior to the Tennessee site. They are incomparable, because Tennessee's advantage over New Mexico on biological impact exceeds the veto threshold, while New Mexico's advantage on socioeconomic factors also exceeds the veto threshold.

Using ELECTRE III software, the following credibility matrix is obtained:

Credibility Matrix	Tennessee	New Mexico	California
Tennessee	1	0	0.846
New Mexico	0	1	0.77
California	0.154	0.19	1

These results are then used to determine outranking relationships. Of the three pairs, the relative differences are:

Tennessee vs. New Mexico   0   New Mexico vs. Tennessee   0
    difference 0

Tennessee vs. California   0.846   California vs. Tennessee   0.154
    difference 0.692

New Mexico vs. California   0.77   California vs. Tennessee   0.19
    difference 0.58

A downward pass begins with identifying the maximum difference, which is 0.692 for Tennessee over California. Each credibility matrix entry pair is then considered in descending order. The first-rank order consists of all alternatives that are not outranked by other alternatives. In this case, both Tennessee and New Mexico outrank California. There is no outranking relationship between Tennessee and New Mexico, so the descending outranking relationship is thus:

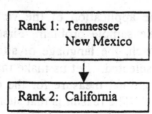

The ascending ranking is the same idea, only identify those alternatives that are outranked. California is outranked by both Tennessee and New Mexico, and thus is assigned the bottom rank. Neither Tennessee nor New Mexico outrank the other, so they are both at the top rank. This yields an upward ranking identical to the downward ranking given above. The final ranking is the average of the two rank-orders (which of course in this case is the same).

The outranking approach is computationally intensive, but flexible in that it allows decision makers to set many parameters. This approach uses measures to identify outranking relationships. The next method also uses the concept of outranking, only using more ordinal input, which is expected to be more reliable.

Verbal Decision Analysis

Verbal Decision Analysis (VDA) was developed for the problem of identifying the best alternative choice from a group. The method PACOM (PAir COMpensation - Larichev and Moshkovich, 1997) is based on pair-wise comparison of alternatives. The alternatives are compared to find relative advantages and disadvantages. After this comparison, the decision-maker separately ranks the disadvantages of each alternative. To elicit the decision-maker's preferences, special reference options are created and presented to the decision-maker. The reference options have the same performance on most criteria, but have countervailing performance on a small number of criteria. The comparisons of the reference options made by the decision-maker are used for the selection of best alternative as follows: if all of the disadvantages of one alternative are found to be less negative than those of the other option, the first option is preferred to the other option.

We demonstrate the application of the verbal decision analysis and PACOM method for the sample problem. First, according to the verbal decision analysis approach, the language of alternatives' measurement in the terms of criteria is selected. Let us make the realistic supposition that experts cannot give a cardinal measure for the criteria "Expected deaths" and "Risk of catastrophe". The reason is that such measurements have to be made under great uncertainty. It is very difficult to give a number for deaths expected during the process of the system construction, transportation of material to the site, and exposure of inhabitants in the vicinity of the site. It is even more difficult to estimate the expected number of years before a catastrophe might expose materials stored in the repository.

What can experts express in a relatively reliable manner? They could compare those pairs of alternatives whose performance on criteria are expressed in qualitative terms based on their previous experience. So, they could give relative qualitative evaluations on such criteria. For the other criteria, possible language for the description of the estimates of alternative performance is as follows:

Criterion	best category	intermediate category	worst category
Aesthetic impact	$4 million	$10 million	$ 18 million
Biological	Small		big
Socioeconomic	low	moderate	high
Construction cost	normal		high
Operating cost	low	moderate	high
Risk	less		more

It is difficult to believe that experts could give exact quantitative estimates of construction cost and operating cost long periods of time prior to the construction of a facility. It is well known that in many real cases such estimates are very inaccurate (Merrow, 1988). It is therefore more reasonable to elicit qualitative estimations for each alternative from experts. The same is true for the evaluation of biological (number of species) and socioeconomic (number of job impacts).

Let us compare now New Mexico and California. The evaluations of two alternatives are given below

Criterion	New Mexico	California		Advantage
Expected deaths	much less	much more	*1*	New Mexico
Aesthetic impact	$4 million	$ 18 million	*4*	New Mexico
Biological	equal	equal		neither
Socioeconomic	low	high	*3*	New Mexico
Construction cost	normal	high	*2*	New Mexico
Operating cost	high *1*	moderate		California
Risk	equal	equal		neither

The decision-maker is first asked to give the ranking for the importance of relative disadvantages of both alternatives. It is known that ranking is a more reliable operation (from a psychological point of view) and less sensitive to human errors than direct statement of weights (Larichev and Moshkovich, 1997). The numbers in italics located by the evaluations represent such a ranking. For example, for this decision-maker, the most serious disadvantage of the California site is expected deaths.

To compare the alternatives, it is necessary to create two pairs of reference options. The first pair consists of the alternative "New Mexico" and option *y1* that has the same measures as "New Mexico" on all criteria except the first (Expected deaths) and last (Operating cost). Option *y1* has the criteria estimations of the "California" alternative. So, the two options in the pair have different evaluations on only two criteria. Results of a number of experiments (Larichev, et al., 1978) demonstrate that people are quite reliable in making comparisons of this type. This information could be presented in the following way:

Criterion	New Mexico	*y1*	Advantage
Expected deaths	**much less**	**much more**	**New Mexico**
Aesthetic impact	$4 million	$4 million	neither
Biological	equal	equal	neither
Socioeconomic	low	low	neither
Construction cost	normal	normal	neither
Operating cost	**high**	**moderate**	*y1*
Risk	equal	equal	neither

Comparing the options in the pair, the decision-maker defines his/her preferences between the worst disadvantage of "New Mexico" and that of "*y1*". The decision-maker selects from three possible answers:

```
1. Option 1 is more preferable than option 2.
2. Option 2 is more preferable than option 1.
3. Options 1 and 2 are equally preferable.
```

Let us suppose that the decision-maker gives the first answer. This means that improving the criterion "Expected deaths" is more important for the decision-maker than lowering Operating costs. It is now necessary to check to what extent the result of this comparison depends on the other (equal) estimations on other criteria.

The second pair of reference options consists from alternative "California" and option $y2$, that differs from California only on the first and last criteria. Again, two options in the pair have different evaluations on only two criteria. They could be presented in the following form:

Criterion	y2	California	Advantage
Expected deaths	**much less**	**much more**	*y2*
Aesthetic impact	$18 million	$18 million	neither
Biological	equal	equal	neither
Socioeconomic	high	high	neither
Construction cost	high	high	neither
Operating cost	**high**	**moderate**	**California**
Risk	equal	equal	neither

The decision-maker is to make the same choice as before. Let us suppose that the decision-maker prefers $y2$. This means that improving "Expected deaths" is more important for the decision-maker than improving Operating costs.

The decision-maker preferences are the same in both reference pairs. This means that the decision-maker comparisons are independent from other (equal) evaluations on the other criteria. Larichev and Moshkovich (1997) proved that, in the case of independence, the results of the comparisons of the reference options can be used for the comparison of the alternatives in the pair. In the case when the preferences in reference pairs are different, the criteria are dependent. For such cases, the verbal decision analysis approach is to change the verbal description of the problem (see details in Larichev, Moshkovich,1997). Based on the first pair-wise comparison, we infer that the New Mexico site is more preferable than the California site. Now, we must analyze the second pair: Tennessee-New Mexico.

Criterion	New Mexico		Tennessee		Advantage
Expected deaths	equal		equal		neither
Aesthetic impact	$4 million		$10 million	*1*	New Mexico
Biological	big	*2*	small		Tennessee
Socioeconomic	low		moderate	*2*	New Mexico
Construction cost	equal		equal		neither
Operating cost	approx.equal		approx. equal		neither
Risk	more	*1*	less		Tennessee

The decision-maker must again rank the disadvantages of both alternatives. An assumed set of ranks are given near estimations above. Following the PACOM method, it is necessary to compare the worst disadvantages of both alternatives and then the second disadvantages (if the number of disadvantages of the alternatives is not equal, the disadvantage of one alternative could be compared with several disadvantages of the second alternative). For elicitation of the decision-maker's preferences and for the independence check, it is necessary to create four reference pairs. Let us give only the first:

Criterion	*y3*	Tennessee	Advantage
Expected deaths	equal	equal	neither
Aesthetic impact	**$4 million**	**$10 million**	*y3*
Biological	small	small	neither
Socioeconomic	moderate	moderate	neither
Construction cost	equal	equal	neither
Operating cost	approx. equal	approx. equal	neither
Risk	**more**	**less**	**Tennessee**

Let us suppose that the decision-maker prefers "Tennessee" in this pair. A second reference pair is needed to check the independence condition. We also need two more reference pairs to compare the estimations of alternatives on the criteria "Biological" and "Socioeconomic". To eliminate the possibility of random errors in decision-maker responses, we could make the additional comparison: compare the first disadvantage of alternative 1 with the second disadvantage of alternative two. Conducting these additional comparisons will provide at least two confirmations of every decision maker answer – referred to as "closed" procedures (see Larichev and Moshkovich, 1997). If the comparisons yield the Tennessee site as preferred again, then the output of the decision process is evident. If the second comparison were to yield a different result, the situation of incomparability arises. In the situation of incomparability, the decision-maker could use different strategies. First, he/she could again analyze the difference in the evaluation for both alternatives. If, for example, the decision-maker could accept that the difference in the evaluation on the Socioeconomic criterion is not really that significant, the Tennessee site is identified as the best alternative.

In the general case, when the decision-maker accepts the estimations of both alternatives, the ASTRIDA method (Berkeley et al., 1991) could be used. To resolve the problem of incomparability, a new, more attractive option is sought that is better than the original two. This approach helps decision-maker find the minimum changes in options' evaluations needed to create a new, better alternative. For example, it is possible to take some additional measures (connected with additional expenses), to improve the evaluation of Tennessee on Socioeconomic criterion. In this case a new alternative - an improved Tennessee, appears with the same evaluation on the Socioeconomic criterion as New Mexico, but with some additional expenses. If such expenses are relatively moderate, the decision-maker could accept improved Tennessee as the best option. If not, he/she could again ask expert to find a new, better option.

Let us make two additional comments. First, in the case of several alternatives, it is not obligatory to make all pair-wise comparisons. The results of psychological studies (Montgomery and Svenson, 1989) demonstrate that people could make the preliminary selection of potentially best alternatives and accurately use these as the basis for pair-wise comparisons. Second, method ASTRIDA cannot itself to create the better alternative. But it can show the direction of the search, and the minimal changes needed in the evaluations of one of the alternatives. Using this direction of change, the decision-maker could ask experts to develop a new, better alternative. Practical demonstration of the efficiency of this approach is given in the Chapter 7. Finally, let us stress that the verbal decision analysis approach is based on a psychologically valid way of eliciting information from decision-makers and experts.

Summary

Four multicriteria methods were demonstrated as alternatives to cost-benefit analysis. If full knowledge is available, and if preferences are accurately expressed in monetary terms, cost-benefit analysis provides the most thorough means of analysis. However, cost estimates are rarely as accurate as the point-estimates used imply. That factor, as well as the need to consider the relative utility of criteria measures, make the use of multiattribute utility analysis attractive. These four multicriteria methods of implementing decision analysis provide a set of tools for different conditions of data availability and accuracy. MAUT is best applied with objective measures, along with carefully measured single-attribute utility functions measuring value for a decision-maker, and carefully elicited tradeoffs among criteria. AHP has been widely used to convert subjective concepts into quantitative value functions, although this would be less accurate than objective measures. The outranking methods require less precise input data, and focus on relative advantages and disadvantages between all pairs of alternatives. Relative weights among criteria form the basis for measuring the degree of preference of one alternative over another. Roy and Bouyssou (1986) compared the outranking method ELECTRE III with MAUT on a nuclear power plant siting example, and examined relative advantages of the outranking approach. Finally, verbal decision analysis was presented. This method requires the least precision in data, focusing rather on categories distinguishing relative levels of performance over the criteria. The cost of improving one alternative in order to make it at least as attractive as the other alternative on all criteria is estimated. This simplifies the decision problem, focusing on significant differences among the alternatives being considered.

References

Barzilai, J., Cook, W.D. and Golany, B. Consistent weights for judgements matrices of the relative importance of alternatives, *Operations Research Letters* 6:3, 1987, 131-134.

Berkeley, D., Humphreys, P., Larichev, O. and Moshkovich, H. Aiding strategic decision making: Derivation and development of ASTRIDA. In *Environment for Supporting Decision Processes*, Y. Vecsenyi and H. Sol, eds., North Holland, Amsterdam, 1991.

Kahneman, D., Slovic, P. and Tversky, A. (eds.), *Judgement Under Uncertainty: Heuristics and Biases*, Cambridge: Cambridge University Press, 1982.

Keeney, R.L., *Siting Energy Facilities*, Academic Press, New York, 1980.

Keeney, R.L. & Raiffa, H., *Decisions with Multiple Objectives*, Wiley, New York, 1976.

Kunreuther, H.C., Linnerooth, J., Lathrop, J., Atz, H., Macgill, S., Mandl, C., Schwarz, M. and Thompson, M. *Risk Analysis and Decision Processes: The Siting of Liquified Energy Gas Facilities in Four Countries*. Springer-Verlag, Berlin, 1983.

Larichev, O.I. and Moshkovich, H.M. *Verbal Decision Analysis for Unstructured Problems*, Kluwer Academic Publishers, Boston, 1997.

Larichev, O.I., Yu, A., Zuev, Y.A. and Gnedenko, L.S. Method ZAPROS (Closed Procedures near Reference Situations) for the analysis of variants of complex decisions, in S.V. Emelyanov, ed., *Multicriteria Choice for the Solution of Ill-Structured Problems*, Moscow: VNIISI Proceedings, N 5 (in Russian), 1978, 83-97.

March, J.G. and Simon, H.A. *Organizations*, New York, Wiley, 1958.

Merrow, E.W.: *Understanding the Outcomes of Megaprojects: A Quantitative Analysis of Very Large Civilian Projects*. Rand Corporation, Santa Monica, CA, 1988.

Montgomery, H. and Svenson, O. A think-aloud study of dominance structuring in decision processes, in *Process and Structure on Human Decision Making*, H. Montgomery and O. Svenson, eds., J. Wiley and Sons, Chichester, 1989, 135-150.

Roy, B. Clessement et choix en presence de criteres multiples. *RIRO* vol. 8, 1968, pp. 57-75.

Roy, B. and Bouyssou, D. Comparison of two decision-aid models applied to a nuclear power plant siting example. *European Journal of Operational Research*, vol. 25, 1986, 200-215.

Saaty, T.L. A scaling method for priorities in hierarchical structures. *Journal of Mathematical Psychology* 15:3 1977, 234-281

Saaty, T.L. *The Analytic Hierarchy Process* New York, McGraw Hill 1980 (2nd ed. T.L. Saaty, 1988).

Saaty, T.L. *Decision Making for Leaders* Belmont, CA Van Nostrand Reinhold 1982 (2nd ed. RWS Publications, 1986).

Sarin, R.K. A social decision analysis of the earthquake safety problem: The case of existing Los Angeles buildings. *Risk Analysis*, vol. 3, no. 1, 1983, 35-50.

Shim, J.P. Bibliographical research on the analytic hierarchy process (AHP). *Socio-Economic Planning Sciences* 23:3 1989, 161-167

Simon, H.A. *The New Science of Management Decision*, New York: Harper and Row Publ., 1960.

Smith, V.K. and Desvousges, W.H. The value of avoiding a LULU: Hazardous waste disposal sites. *The Review of Economics and Statistics*, vol. 68, no. 2, 1986, 293-299.

Walker, W.E., Abrahamse, A., Bolten, J., Kahan, J.P., Van de Riet, O., Kok, M., and Den Braber, M. A policy analysis of Dutch river dike improvements: Trading off safety, cost, and environmental impacts. *Operations Research* vol. 42, no. 5, 1994, 823-836.

Zahedi, F. The analytic hierarchy process - A survey of the method and its application., *Interfaces* 16 (4) 1986, 96-108

3 HIGH LEVEL WASTE REPOSITORY SELECTION

Of the many multiple criteria studies, probably the most well-known is the study conducted by Woodward-Clyde Consultants concerning the location of a permanent storage facility for nuclear waste. Nuclear waste from power plants in the US is a major problem. In 1982, the Nuclear Waste Policy Act was passed by the US Congress, specifying a schedule for selecting repository sites. In November, 1984 the Department of Energy established general guidelines for evaluation of possible repository sites, specifying system guidelines for public health and safety, the environment, socioeconomics, and the ease and cost of repository development, as well as technical guidelines including conditions required in such a site. The repository was designed to be a system of tunnels and rooms excavated in stable rock at least 1000 feet underground. The plan was for nuclear waste to be shipped from power plants to the repository and permanently stored there. When full, the repository is to be sealed to minimize radiological leakage. (Keeney, 1987, p. 196). In December, 1984 the Department of Energy published initial environmental assessments of five sites nominated by the Secretary of Energy (DOE, 1984a). These sites were Davis Canyon, Utah (bedded salt); Deaf Smith, Texas (bedded salt); Richton Dome, Mississippi (salt dome); Hanford, Washington (basalt); and Yucca Mountain, Nevada (volcanic tuff). The original sites selected for characterization in 1984 were Hanford, Washington, Yucca Mountain, Nevada, and Deaf Smith County, Texas, based on rankings of the five sites against the technical guidelines (DOE, 1984b).

This selection process was reviewed by a number of agencies, including the National Academy of Sciences, who called the evaluation procedure unsatisfactory and biased, among other things. The Department of Energy therefore conducted a more thorough evaluation using multiattribute utility analysis (Keeney and Raiffa, 1976) along with an independent study by the Board on Radioactive Waste Management of the National Academy of Sciences. This process led to a 28 May, 1986 announcement by the Department of Energy that they were recommending Yucca Mountain, Nevada; Deaf Smith County, Texas; and Hanford, Washington for characterization, or detailed study (DOE, 1986).

This project has been very widely reported in the decision-making literature. Merkhofer and Keeney (1987) provided a very thorough review of the multiattribute utility analysis conducted. Gregory and Lichtenstein (1987) reviewed the process as of that time, which had yielded the 1986 recommendation to conduct detailed study of Yucca Mountain (ranked first by the multiattribute study), Deaf Smith County (ranked second by the multiattribute study), and Hanford (ranked fifth by the multiattribute study. Keeney (1987) questioned the selection of these three specific sites in light of the high cost of characterization (at least $1 billion per site). Later studies related to the Yucca Mountain site include analysis of the transportation of nuclear waste (Rautman, Reid and Ryder, 1993), documenting the probabilistically centered elicitation of expert evaluations of future events (DeWispelare, Herren and Clemen 1995), and a reassessment of the entire repository strategy (Keeney and von Winterfeldt, 1994).

Multiattribute Utility Analysis

The multiattribute utility analysis consisted of five basic steps (Merkhofer and Keeney, 1987):

1. Establish objectives and develop performance measures to quantify the degree to which each site meets each objective.

2. Specify a set of postclosure scenarios that might effect the performance of the repository system if they occurred.
3. Quantify estimated preclosure performance of each site. For each postclosure scenario, quantify estimated postclosure performance.
4. Develop a quantitative model of values to combine various performance measures into an overall measure of desirability.
5. Calculation values for each site, and perform sensitivity analysis to determine which technical judgments and value judgments are most critical.

Three groups of analysts were created. Selected Department of Energy staff and decision analysts were charged with the analysis. A second group was formed to provide technical judgments, and a third group to obtain value judgments. The technical specialists were selected for their knowledge of the five specific environmental assessments and siting guidelines. The technical group developed measures to quantify site performances on criteria. The list of criteria measures was developed by the decision analysts along with the technical specialists, as were the scenarios for postclosure performance and judgmental estimates of each site's performance on each measure. Value judgments were provided independent of technical judgements by four senior managers from the Office of Civilian Radioactive Waste Management.

Objective Development

A hierarchy of preclosure objectives was developed, with four major hierarchical elements. This preclosure hierarchy is shown in Table 1.

Table 1: Hierarchy of Objectives – Nuclear Waste Repository Siting Study (Merkhofer & Keeney, 1987)

```
Overall Objective
Major Elements
        Second Level
                Third Level
                        Fourth Level
        measure

Minimize Adverse Preclosure Impacts
        Health & Safety
                At the Repository
                        Radiological Health Effects
                                Minimize Worker Health Effects    X₁
                                Minimize Public Health Effects    X₂
                        Nonradiological Health Effects
                                Minimize Worker Health Effects    X₃
                                Minimize Public Health Effects    X₄
                        Radiological Health Effects
                In Transportation
                                Minimize Worker Health Effects    X₅
                                Minimize Public Health Effects    X₆
                        Nonradiological Health Effects
                                Minimize Worker Health Effects    X₇
                                Minimize Public Health Effects    X₈
Environmental
        Minimize Aesthetic Degradation                            X₉
        Min Archaeological, Historical, & Cultural Degradation     X₁₀
        Minimize Biological Degradation                           X₁₁
        Minimize Adverse Socioeconomic Impacts                    X₁₂
Economic Costs
        Minimize Repository Costs                                 X₁₃
        Minimize Waste Transportation Costs                       X₁₄

Minimize Adverse Postclosure Impacts
        First 10,000 years
                Minimize cumulative releases of EPA standard      Y₁
        Years 10,000 to 100,000
                Minimize cumulative releases of EPA standard      Y₂
```

Table 2 gives the fourteen pre-closure measures as well as two post-closure measures.

Table 2: Performance Measures and Units
(Merkhofer and Keeney, 1987)

Measure	Units
Preclosure objectives	
X_1 Repository-worker radiological fatalities	Number of cancer deaths
X_2 Public radiological fatalities from repository	Number of cancer deaths
X_3 Repository-worker nonradiological fatalities	Number of accident deaths
X_4 Public nonradiological fatalities from repository	Number of accident deaths
X_5 Transportation-worker radiological fatalities	Number of cancer deaths
X_6 Public radiological fatalities	Number of cancer deaths
X_7 Transportation-worker nonradiological fatalities	Number of accident deaths
X_8 Public nonradiological fatalities from transport	Number of accident deaths
X_9 Adverse aesthetic impacts	scale ranging from 0 to 6
X_{10} Adverse archaeological, historical, & cultural	scale ranging from 0 to 5
X_{11} Adverse biological degradation	scale ranging from 0 to 5
X_{12} Adverse socioeconomic impacts	scale ranging from 0 to 4
X_{13} Repository costs	millions of dollars
X_{14} Waste-transportation costs	millions of dollars
Postclosure objectives	
Y_1 Cumulative releases of radionuclides 10k years	multiples of EPA standard
Y_2 Cum releases of radionuclides 10k to 100k yrs	multiples of EPA standard

Risk Scenarios

As part of the risk analysis, a series of fourteen significant scenarios were hypothesized. These are given in Table 3.

Table 3: Potentially Significant Scenarios
(Merkhofer and Keeney, 1987, p. 180)

Scenario	Description
1	Expected conditions based on available information
2	Unexpected features
3	Repository-induced dissolution of the host rock
4	Advance of a dissolution front
5	Movement on a large fault inside the controlled area but outside the repository
6	Movement on a large fault within the repository
7	Movement on a small fault inside the controlled area but outside the repository
8	Movement on a small fault within the repository
9	Movement on a large fault outside the controlled area
10	Extrusive magmatic event occurring during the first 500 years after closure
	Extrusive magmatic event occurring 500 to 10000 years after closure
11	Intrusive magmatic event
12	Large-scale exploratory drilling
13	Small-scale exploratory drilling
14	Incomplete sealing of the shafts and the repository

Probability assessment procedures were outlined in Keeney and von Winterfeldt (1991).

Performance Measures

Data, based on technical studies, was provided for each of the five sites on the fourteen pre-closure measures. These are given in Table 4.

Table 4: Measures of Preclosure Impacts (Merkhofer and Keeney, 1987)

Measure	Davis Canyon	Deaf Smith	Richton Dome	Hanford	Yucca Mountain	K_I
X_1 worker-rad	2 (1-4)	2 (1-4)	2 (1-4)	9 (2-17)	4 (1-9)	1
X_2 public-rad	< 0.1 (<0.1-0.2)	0.5 (0.1-1)	0.7 (0.3-1.5)	0.7 (<0.1-1.5)	0.1 (<0.1-0.1)	4
X_3 worker-acc	27 (17-36)	29 (19-39)	27 (17-36)	43 (28-58)	18 (12-24)	1
X_4 public-acc	0 (0-0)	0 (0-0)	0 (0-0)	0 (0-0)	0 (0-0)	4
X_5 trans-rad	0.73 (0-1.0)	0.64 (0-0.9)	0.52 (0.0.73)	0.9 (0-1.3)	0.81 (0-1.1)	1
X_6 public-rad	3.5 (0-4.9)	2.9 (0-4.1)	2.4 (0-3.4)	4.3 (0-6.1)	4.1 (0-5.7)	4
X_7 worker-trans	2.1 (0.96-0.34)	1.6 (0.73-2.6)	1.3 (0.6-2.1)	2.7 (1.2-4.3)	2.5 (1.1-4.0)	1
X_8 public-trans	8.4 (3.9-13.5)	6.7 (3.1-10.8)	5.3 (2.4-8.5)	11.0 (5-17.7)	10.21.98 (4.7-16.4)	4
X_9 aesthetic	6 (6-6)	4 (3-5)	4 (1-5)	1 (1-2)	4 (1-5)	1
X_{10} arch,hist	3 (2.5-5)	1 (0-2.5)	0.5 (0-1)	0.5 (0.5-3)	2 (2-3.5)	0.2
X_{11} biological	3.5 (2.67-4.5)	2.33 (1.5-3)	2.67 (2-3.5)	2.33 (1-3.5)	2 (1-2.67)	0.3
X_{12} socioecon	2 (1.33-3)	1.67 (1-3)	2 (1-3)	0.33 (0-0.67)	0.67 (0.33-2)	5
X_{13} rep costs	10,400 (6760-14040)	9,500 (6175-12825)	9,000 (5850-12150)	12,900 (8385-17415)	7,500 (4875-10125)	1
X_{14} trans costs	1,240 (330-2600)	1,120 (300-2350)	970 (260-2040)	1,450 (390-3040)	1,400 380-2940)	1
Y_1 early rad-rel	0.0001 (<0.0001-0.001)	0.0001 (<0.0001-0.001)	0.0001 (<0.0001-0.001)	0.001 (0.001-1)	0.0001 (<0.0001-0.03)	0.53
Y_2 late rad-rel	0.001 (<0.001-0.01)	0.003 (0.001-0.03)	0.001 (<0.001-0.01)	0.03 (0.001-0.1)	0.003 (0.001-0.3)	0.47

(ranges in parentheses)

Multiattribute Utility Function

Impact estimates on each of the measures were aggregated into an overall measure of site desirability through a multiattribute utility function. Independence checks were conducted, and these checks indicated that the measures were independent. Therefore, an additive utility function was used, expressing the value of each alternative as:

$$U(\text{alternative j}) = k_{pre} \, U_{pre}(\text{alternative j on 14 pre-measures})$$
$$+ k_{post} \, U_{post} \, (\text{alternative j on 2 post-measures})$$

The positive scaling factors k_{pre} and k_{post} summed to 1.0. They reflect the tradeoff in changes on pre- and post- measures. The utility functions U, U_{pre} and U_{post} are scaled from 0 to 100. The preclosure utility function was:

where K_i is a positive scaling factor representing value tradeoffs between units of performance measures, C_i terms represent the disutility functions for the fourteen performance measures, and the constants 121 and 0.05 scaled preclosure utility to a scale ranging from 0 to 100. The fourteen K_i

$$U_{pre}(\text{alternative } j) = 121 - 0.05 \sum_{i=1}^{14} K_i C_i(x_{ij})$$

are included in the last column of Table 4. These were developed from value judgments elicited from four senior managers of the Office of Civilian Radioactive Waste Management within the U.S. Department of Energy. Disutility functions C_i (with higher scores worse than lower) were the values in Table 4 for measures 1 through 8. For the four scaled measures (9 through 12), the C_i values are shown in Table 5. The C_i for costs (measures 13 and 14) were the dollar values in millions.

Table 5: Scales and utility equivalents for environmental and
socioeconomic factors

X_9 Adverse aesthetic impacts	scale	C_i
none	0	0
one minor effect	1	3
two minor effects	2	6
three minor effects	3	9
one major effect	4	33
two major effects	5	67
three major effects	6	100

X_{10} Adverse archaeological, historical, & cultural impacts

	scale	C_i
No impacts on any significant historical properties	0	0
Min adverse impact to 1 major or 5 minor historical	1	12
Min adverse impact to 2 major or 10-minor historical	2	23
Major adverse impact to 2 major or 10-minor historical	3	56
Major adverse impact to 3 major or 15-minor historical	4	78
Major adverse impact to 4 major or 20-minor historical	5	100

(by criteria of National Register of Historic Places;
minor significance has local significance;
major impacts change integrity or significance of property;
minor impacts may alter the property but not change integrity or significance.)

X_{11} Adverse biological impacts

	scale	C_i
No damage to plants or wildlife or habitats	0	0
Damage to species or habitats common in region	1	4
Damage to species not threatening regional abund.	2	10
Damage to species not threatening regional abund.	3	18
Damage threatens regional abundance of species	4	40
Damage threatens regional abundance of species	5	100

X_{12} Adverse socioeconomic impacts

	scale	C_i
No social or economic disruptions, or water impact	0	0
In-migrat. pop. of 5,000, min. upgrading, no water imp.	1	8
In-migrat. pop. of 5,000, major upgrade, social disrupt.	2	20
In-migrat. pop. of 10,000, major upgrade, social disrupt.	3	60
In-migrat. pop. of 10,000, major upgrade, major water.	4	100

The postclosure utility function used was:

$$U_{post}\ (\text{alternative } j) = 100[k_1\ U_1(Y_1) + k_2\ U_2(Y_2)]$$

where k_1 and k_2 are scaling constants reflecting tradeoffs of performance between the shorter period (the first 10,000 years) and the longer period (10,000 to 100,000 years). Cumulative releases of radionuclides were expressed as a fraction of Environmental Protection Agency limits. The utility function $U_1 = 1 - Y_1$, while $U_2 = 1 - Y_2/9$. Tradeoff values were $k_1 = 0.53$ and $k_2 = 0.47$.

Utility Analysis Results

The utility functions were used to identify the relative performances of the five sites. Table 6 provides aggregates for expected lives lost and costs, as well as aggregate measures reflecting value tradeoffs.

Table 6: Aggregate Performances of Alternative Sites

	Davis Canyon	Deaf Smith	Richton Dome	Hanford	Yucca Mountain
Expected Fatalities	44	43	39	72	40
Weighted Fatalities	80	74	64	120	83
Health & Safety (X_1-X_8)	80	74	64	120	83
Environ & Socio (X_9-X_{12})	220	119	139	23	71
Expected Costs $(X_{13}-X_{14})$	11,640	10,620	9,970	14,350	8,900
Preclosure Utility	61.3	66.9	70.1	48.5	75.7
Postclosure Utility	99.99	99.98	99.99	99.76	99.98
Rank-Health & Safety	3	2	1	5	4
Rank-Environ & Socio	5	3	4	1	2
Rank-Expected Costs	4	3	2	5	1
Rank-Preclosure	4	3	2	5	1
Rank-Postclosure	1	3	2	5	4
Rank-Combined	4	3	2	5	1

(weighted fatalities gave a weight of 4 to public fatalities, and 1 to worker fatalities)

Sensitivity Analysis

Different criteria led to radically different rankings.

Alternative Rankings – High Level Nuclear Waste Repository		
Public Health & Safety	Environment, Socioeconomics	Cost
1 Richton Dome	Hanford	Yucca Mountain
2 Deaf Smith County	Yucca Mountain	Richton Dome
3 Davis	Deaf Smith County	Deaf Smith County
4 Yucca Mountain	Richton Dome	Davis
5 Hanford	Davis	Hanford

Differences in site utility function was attributed largely to differences in transportation requirements. Based on health and safety factors, the ranking of sites was 1-Richton Dome, 2-Deaf Smith, 3-Davis Canyon, 4-Yucca Mountain, and 5-Hanford. The ranking based on environmental and socioeconomic factors was 1-Hanford, 2-Yucca Mountain, 3-Deaf Smith, 4-Richton Dome, and 5-Davis Canyon. The expected health and safety impact, as well as environmental and socioeconomic impact, was expected to be quite low (as the site selection was screened to consider locations with attractive ratings on these factors). Cost therefore had a much higher weight, and led to the ranking 1-Yucca Mountain, 2-Richton Dome, 3-Deaf Smith, 4-Davis Canyon, and 5-Hanford. Merkhofer and Keeney (1987) considered a number of rankings reflecting various emphases. Calculations were rechecked by those authors to consider all reasonable combinations of impacts within estimated ranges of uncertainty for all value judgments obtained from DOE managers, as well as for a variety of attitudes of risk. The same final ranking was obtained.

Sensitivity analysis was also applied to postclosure utility changes due to estimated releases under various scenarios (from Table 3), along with probabilities of each scenario occurring, time intervals, and parameters of the utility model. This sensitivity analysis indicated that the most important assumptions were the scenario release estimates and scenario probabilities.

Therefore, 98 percent confidence ranges for postclosure utilities were generated
Expert judgment elicitation played an important role in this study.

The Merkhofer and Keeney (1987) analysis indicated that all five sites were expected to provide an exceptionally high degree of waste isolation for at least 100,000 years after repository closure. The Hanford site was less favorable than the other four sites, but still far above the threshold of acceptability established by the Environmental Protection Agency. Confidence in the three salt sites was extremely high, better than that of the nonsalt sites (Yucca Mountain and Hanford). Overall, the base-case ranking shown in Table 6 was found to be relatively insensitive to uncertainties or value judgments.

The relative importances assigned to preclosure and postclosure scaling factors k_{pre} and k_{post} were found to have little impact on the final ranking of the five alternative sites. When nearly all weights was assigned to postclosure, the final ranking changed to the postclosure utility ranking given in Table 6, yielding Davis Canyon as the preferred site be a very bare margin, with only Hanford noticeably lower in value. Otherwise, the ranking had Yucca Mountain as the preferred site, as indicated in the final ranking given in Table 6. Sensitivity analysis considering use of optimistic assumptions, pessimistic assumptions, and combinations of these for preclosure and postclosure parameters did yield changes in site ranking depending on the weights assigned to preclosure and postclosure. But the Hanford site was always ranked last. Among the salt sites, Richton Dome was always ranked higher than Deaf Smith, which in turn always ranked higher than Davis Canyon. The Yucca Mountain site was the most sensitive to pessimistic, base-case, or optimistic assumptions. For pessimistic assumptions of postclosure performance, Yucca Mountain had lower utility due to the possibility of large radionuclide releases in the unexpected features scenario. For pessimistic assumptions of postclosure performance, Yucca Mountain was ranked among the top three sites only when k_{post} was less than roughly 0.35. For base-case or optimistic assumptions, the Yucca Mountain site was almost always ranked first for all weights for preclosure and postclosure factors.

Study Reviews

The Board on Radioactive Waste Management of the National Academy of Sciences reviewed the analytic techniques applied in March, 1986. Because this Board did not want to address ultimate ranking or specific sites, it limited its review to one preclosure site analysis, along with all five postclosure site analyses. Data and judgments were not reviewed in detail. The Board commended the Department of Energy for the high quality of the study. The multiattribute utility method was considered appropriate, and the care and attention to detail with which it was applied was noted. The multiattribute utility method was lauded for stating assumptions, judgments, preferences, and tradeoffs clearly and systematically.

Gregory and Lichtenstein (1987) applauded the multiattribute utility analysis because its assumptions were explicitly stated, the method provided a basis for comparing unlike attributes through explicit value tradeoffs, a concrete number was obtained for output allowing unambiguous ranking, uncertainty was easily incorporated, and sensitivity analysis could be conducted. While commending the multiattribute analysis as extremely well done, Gregory and Lichtenstein raised concerns about the reliance on expert assessments of probabilities of events 100,000 years into the future. The choice of attributes was mandated by guidelines, but some concerns about them were expressed. Health and safety impacts were entirely measured by expected deaths, which were expected to be sufficiently correlated with nonfatal health effects so that the result would not have been different had nonfatalities been included. Gregory and Lichtenstein felt that leaving out nonfatal health effects might have led to underestimation of the value of health effects relative to other attributes, which were more thoroughly described. With respect to socioeconomic impacts, potentially important factors such as impact on tourism and business development due to public fear were not included. But in general, Gregory and Lichtenstein were very positive about the study.

Results

Secretary of Energy Herrington announced that Yucca Mountain, Deaf Smith, and Hanford had been selected for characterization (U.S. Department of Energy, 1986; Keeney, 1987). Characterization involves detailed assessment of project activity cost and compliance with requirements, as well as construction of an exploratory shaft, subsurface excavation and tunneling, and testing of host rock. The multiattribute analysis was intended in the spirit of a decision support system, with the understanding that the decision would be made by the Department of Energy, possibly differing from the analysis. One factor is that the Department of Energy had a portfolio selection decision, where three sites were chosen, not one. Therefore, diversity of sources would be attractive. Another factor was that value tradeoffs were obtained from four DOE managers. The final decision was made by the Secretary of Energy, who might share many values with the four managers used for tradeoff assessment, but some differences were certainly possible. A number of criteria could have been added to the study, including less-than-fatal health effects, and critically important in the outcome of the case, local opposition to the repository. However, the Department of Energy stated that the study provided significant insights about the relative desirability of the sites.

The repository program is at a virtual impasse, as the State of Nevada has taken a strong position against the Yucca Mountain site, and the Department of Energy continues to study the matter in depth. One view of the local opposition generated by the Yucca Mountain proposal is presented by Ginsburg (1994). Kunreuther and Easterling (1990) reported the results of a study seeking to identify the level of compensation that would be required for Nevada voters to accept the repository. They found no statistical evidence that preference for the repository differed as an annual rebate was increased. Risk appeared to be inherently noncompensable. Kunreuther and Easterling concluded that compensation in the form of a rebate would be unlikely to have a positive effect. It appeared to those researchers that benefits were simply rejected out of hand unless the safety of the facility and integrity of the siting process were assured.

The Department of Energy postponed the decision to open the Yucca Mountain site until at least 2010 (Wald, 1989) after spending $500 million on study (Kunreuther and Easterling, 1990). This is primarily because of strong opposition to the site on the part of the citizens of the State of Nevada, who have elected officials who ran for office on the platform of opposing use of the site. (Current congressional targets move this back to 2007.)

Keeney (1994) later reviewed cost escalation of the study of nuclear waste disposal.

Cost Estimates for:	characterization	construction	open date
1982	$35 million	$1 billion	1998
1986	$1 billion	$8.9 billion +	2003
1989	$11.5 billion	$14.2 billion	2010

Keeney and von Winterfeldt (1993) concluded that the use of Yucca Mountain as a repository for permanent storage should be stopped, and that the Nuclear Waste Policy Act should be amended to include more reasonable strategies for managing nuclear waste. They also concluded that an independent detailed analysis be adopted to design a better national policy.

References

DeWispelare, A.R., Herren, L.T. & Clemen, R.T. "The Use of Probability Elicitation in the High-Level Nuclear Waste Regulation Program," *International Journal of Forecasting*, vol. 11, no. L1, 1995, 5-24

Ginsburg, S. *Nuclear Waste Disposal: Gambling on Yucca Mountain*, Aegean Park Press, 1994.

Gregory, R. & Lichtenstein, S. "A Review of the High-Level Nuclear Waste Repository Siting Analysis," *Risk Analysis* 7:2, 1987, 219-223

Keeney, R.L. "An Analysis of the Portfolio of Sites to Characterize for Selecting a Nuclear Repository," *Risk Analysis* 7:2, 1987, 195-218

Keeney, R.L. "Structuring Objectives for Problems of Public Interest," *Operations Research*, vol. 36, no. 3, 1988, 396-405

Keeney, R.L. "Using Values in Operations Research," *Operations Research*, vol. 42, no. 5, 1994, 793-813

Keeney, R.L. & Raiffa, H., *Decisions with Multiple Objectives*, Wiley, New York, 1976.

Keeney, R.L. & von Winterfeldt, D. "Managing Nuclear Waste from Power Plants," *Risk Analysis* 14, 1994, 107-130

Kunreuther, H. and Easterling, D. Are risk-benefit tradeoffs possible in siting hazardous facilities? *The American Economic Review*, vol. 80, no. 2, Papers and Proceedings of the Hundred and Second Annual

Meeting of the American Economic Association (May, 1990), 1990, 252-256.

Merkhofer, M.W. & Keeney, R.L. "A Multiattribute Utility Analysis of Alternative Sites for the Disposal of Nuclear Waste," *Risk Analysis*, vol. 7, no. 2, 1987, 173-194

Rautman, C.R., Reid, R.A. & Ryder, E.E. "Scheduling the Disposal of Nuclear Waste Material in a Geologic Repository Using the Transportation Model," *Operations Research*, vol. 41, no. 3, 1993, 459-469

U.S. Department of Energy, *A Multiattribute Utility Analysis of Sites Nominated for Characterization for the First Radioactive-Waste Depository – A Decision-Aiding Methodology*, DOE/RW-0074, Office of Civilian Radioactive Waste Management, Washington, 1986.

U.S. Department of Energy, *General Guidelines for the Recommendation of Sites for the Nuclear Waste Repositories*, 10 CFR Part 960, Federal Register 49, No. 286, pp. 47714-47770, 1984.

U.S. Department of Energy, *Recommendation by the Secretary of Energy of Candidate Sites for Site Characterization for the First Radioactive-Waste Repository*, DOE/S-0048, Office of Civilian Radioactive Waste Management, Washington, D.C., 1984.

Wald, M.L. U.S. will start over on planning for Nevada nuclear waste dump. *New York Times*, November 29, 1989.

4 ANALYSIS OF ALTERNATIVE METHODS TO DISPOSE OF PLUTONIUM

Strategic Arms Reduction Treaties between the Soviet Union and the United States led to the need to dismantle many nuclear weapons and to dispose of the nuclear material to minimize security risks. This trend was magnified after the dissolution of the Soviet Union because Russia and the United States have undertaken even deeper reductions in tactical nuclear weapons under reciprocal unilateral initiatives. There are between 10 to 20 thousand excess warheads in the U.S., and at least a similar number in the former Soviet Union, containing well over 100 metric tons of plutonium (Holdren, et al., 1995), one of the most toxic and persistent materials known. Major concerns are to minimize the risk that plutonium could be used by terrorists, to minimize the risk that plutonium could be reintroduced into arsenals, and to strengthen arms control to assure continued arms reductions and to prevent the spread of nuclear weapons.

There are about 50,000 tons of surplus plutonium in the U.S. expected to become excess and available for disposition by the year 2003 (Garwin, 1996). This material is in the form of warheads, as well as material at warhead assembly plants, in breeder reactors, and in the form of contaminated waste. The material is to be processed to the degree that it is no more useful than spent nuclear fuel (the "spent fuel standard"). Plutonium can be used in nuclear reactors to generate electricity, but is not as efficient as other fuel forms.

In September, 1993 President Clinton announced that where possible, stockpiles of plutonium would be eliminated, ensuring that they are subject to the highest standards of safety, security, and international accountability. The Office of Fissile Materials Disposition (OFMD), as part of the Department of Energy (DOE), began a comprehensive review of long-term options for plutonium disposition (DOE-SCR 95). That study (phase I) considered 36 disposal options, using nine screening criteria.

Phase I Study

The primary objective was to counter nuclear proliferation to additional states and to terrorist groups. A number of approaches were considered as means to reduce or prevent access to excess weapon materials, listed in Table 1. Alternative approaches were analyzed for compliance with required features, such as treaties, as well as for technical feasibility.

Table 1: Plutonium Disposition Screening Process (OFMD [1995])

STORAGE OPTIONS		
s1	no disposition action (continued long-term storage)	baseline
s2	radiation barrier alloy (storage)	eliminated (es&h)
DIRECT DISPOSAL OPTIONS		
d1	direct emplacement in high-level waste repository	disqual (retrievable, time)
d2	**deep borehole (immobilized)**	**RETAINED**
d3	**deep borehole (direct emplacement)**	**RETAINED**
d4	discard to waste isolation pilot plant	disqualified (capacity)
d5	hydraulic fracturing	disqual. (tech. viability)
d6	deep well injection	disqualified (es&h)
d7	injection into continental magma	elim. (tech.viab., es&h)
d8	melting in crystalline rock	disqual. (tech. viability)
d9	disposal under ice caps	disqual (tech.viab., es&h)
d10	seabed (placement on ocean floor)	disqual. (es&h, treaty)
d11	sub-seabed emplacement	elim. (tech. viab, institut.)
d12	ocean dilution	disqual. (es&h, treaty)
d13	deep space launch	elim. (retrievable, es&h)
IMMOBILIZATION OPTIONS		
i1	underground nuclear detonation	disqual. (es&h, licensing)
i2	borosilicate glass immobilization (dwpf)	eliminated (es&h, cost)
i3	**borosilicate glass immobilization (new facility)**	**RETAINED**
i4	**ceramic immobilization**	**RETAINED**
i5	metal immobilization	to be determined
i6	**borosilicate glass oxidation/dissolution**	**RETAINED**
REACTOR AND ACCELERATOR OPTIONS		
r1	**Euratom mox fabrication/reactor burning**	**RETAINED**
r2	**existing light water reactors**	**RETAINED**
r2a	**partially completed light water reactors**	**RETAINED**
r3	**evolutionary or advanced light water reactors**	**RETAINED**
r4	naval propulsion reactors	disqual. (transparency)
r5	modular helium reactors	elim. (technical maturity)
r6	**CANDU heavy water reactors**	**RETAINED**
r7	ALMRS with pyroprocessing	elim. (technical maturity)
r8	accelerator conversion/molten salt	elim. (technical maturity)
r9	accelerator conversion/particle bed	elim. (technical maturity)
r10	existing light water reactor with reprocessing	disqual. (theft/diversion)
r11	advanced light water reactor with reprocessing	disqual. (theft/diversion)
r12	accelerator-driven mhrs	elim. (technical maturity)
r13	ALMRS with recycle	disqual. (tech. maturity)
r14	particle bed reactors	elim. (technical maturity)
r15	molten salt reactors	elim. (technical maturity)

Time was a major factor of consideration because accounting and security have to be provided while the material exists in its current form (Garwin, 1996). Further, hazard control during transportation was considered. Garwin reviewed the features of the major alternative approaches.

Sub-seabed disposal would involve placing multi-ton canisters containing plutonium as penetrating projectiles in quiet mud 5 kilometers deep in the ocean. Mud is acoustically transparent, and entrepreneurs would be able to locate and retrieve the canisters. This option also violates the London Dumping Convention, so sub-seabed disposal was eliminated.

Ocean dilution (a method not included in Table 1) would involve injection of a solution containing plutonium from a submarine or surface ship at a depth of 3 to 4 kilometers. This method is simple, fast, and would yield a radiological hazard to sea-life acceptable under current rules. However, this method is illegal under the London Dumping Convention, and was disqualified.

Space disposal would involve launching a craft containing plutonium into orbit. This method was estimated to cost about $24 billion, far more than the estimated cost of the other viable options. Furthermore, this option was not acceptably reliable. The more complete disposal from deep-space launch, or launch into the sun, would cost even more. This option was eliminated due to concerns about retrieval ability, as well as environmental, health, and safety considerations.

Underground explosions were proposed by Russian nuclear weapon laboratories. This would involve a large number of warheads exploded at each event. This approach was viewed by critics as creating a plutonium mine, and was disqualified.

The options retained for further study included vitrification and use of reactors, the methods recommended by the National Academy of Sciences. Borehole options were also considered in the phase II study. The disposition process for all three of these methods begins with the transportation of the material to an oxidation site, where plutonium is transformed into plutonium oxide (PuOx).

Vitrification facilities at Savannah River Laboratory were in operation to dispose of highly radioactive fission product wastes from plutonium production at that site. Vitrification involves application of radionuclides, and encasing this processed material in a matrix. The radionuclides make recovery very difficult chemically, while the matrix makes it problematic to physically recover the material. Vitrification is accomplished through borosilicate glass logs of about 2 tons which are placed in stainless steel canisters about 3 meters in height, 60 centimeters in diameter. A tree-structure of suspended containers would be encased in the glass. This approach was originally estimated to take about 8 years to vitrify 50 tons of plutonium. Concerns with this approach involve potential criticality in the melter, and problems with the Defense Waste Processing Facility being behind schedule. Additional permanent storage must be provided for residual material.

There are about 100 light water **reactors** in the United States. Because of mandates promulgated during the Carter administration, these reactors do not use MOX fuel, which is the most commonly used in other countries such as Canada and France. The complex physics involved in various reactor types led to consideration of a variety of reactors. The process would involve oxidation of plutonium (possibly at a different site) and burning as fuel in reactors. Significant radioactive waste would be left over, requiring processing for final deposit. Additional permanent storage must be provided for residual material.

Disposal in deep **boreholes** would involve placing concentrated plutonium in capsules in boreholes 3 to 4 kilometers deep. Material would be barred from terrorists, but retrievable by the host government. It is this last factor that caused resistance to this option on the part of the Russian Government.

While vitrification was the preferred approach of the National Academy of Sciences study, a parallel reactor program was recommended due to concerns listed in the vitrification paragraph (Holdren, et al., 1995).

On January 14, 1997, the Department of Energy released its record of decision recommending a hybrid strategy (dual deployment of a reactor and immobilization alternative) (DOE-ROD 97). In early 1995, the Office of Fissile Material Disposition initiated an independent evaluation of these alternatives conducted by scientists from the Amarillo National Resource Center for Plutonium (ANRCP) and Lawrence Livermore National Laboratories (Dyer, et.al. 1997). The ANRCP team worked closely with OFMD and technical experts during the evaluation process and adopted multiattribute utility theory as a means of assembling the results of detailed technical, economic, schedule, environment, and nonproliferation analyses.

Phase II Study

Eleven options were retained for more detailed analysis (Dyer, et al., 1998). Additional variants were developed to yield improved solutions, and some clearly inferior alternatives were dropped, resulting in the 13 options given in Table 2.

Table 2: Plutonium Disposition Options Considered in Phase II

DIRECT DISPOSAL ALTERNATIVES	
B1	deep borehole with direct emplacement
B2	deep borehole after immobilization
VITRIFICATION ALTERNATIVES	
I1	vitrification at a new facility
I2	vitrification by mixing in glass separated in canister
I3	vitrification in a supplemental melter facility
I4	ceramic vitrification at a new facility
I5	ceramic vitrification separated in canister
I6	electrometallurgical treatment
REACTOR ALTERNATIVES	
R1	existing light water reactors, existing MOX fabrication facilities
R2	partially completed light water reactors operated by DOE
R3	existing light water reactors, new facilities for pit disassembly and MOX fabrication
R4	evolutionary light water reactors (new construction)
R5	CANDU reactors

Phase II of this study focused on evaluation of these alternatives. This stage involved generation of a hierarchy of criteria and measures, followed by more precise estimates of expected operating and performance characteristics of disposal methods, and specification of models to aggregate these measures into an overall score that indicated how well an alternative satisfied the program's objectives. This hierarchy was generated dynamically throughout 1995 by the Office of Fissile Materials Disposition with the guidance of Professor Dyer of the University of Texas. The three overall objectives were nonproliferation, operational effectiveness, and environment safety and health. Each of these was decomposed into operational concepts, which were further represented by specific measures. These criteria are given in Table 3. Adjustments continued into early 1996 to more accurately reflect attainment of objectives.

Table 3: Hierarchy of Criteria for Phase II Analysis

Objective Operational Concept		Measure
Nonproliferation		
Resistance to theft	Material characteristics	DOE attractiveness
		Other separated matls.
	Disposition environment	Bulk throughput
		Number process steps
		Max Pu inventory
		Number of trips
		SST transport miles
	Safeguards & security	Measure uncertainty
		Type of NAS
		Accessibility of Pu
		US classification
Resistance to divers.	Material characteristics	IAEA attractiveness
	Disposition environment	Bulk throughput
		Number process steps
		Max Pu inventory
	Safeguards & security	Measure uncertainty
		Type of NAS
		Accessibility of Pu
		International class.
Irreversibility of mat'l	Attractiveness	Material form
	Accessibility	Material location
International Cooperation		Russian cooperation
		Civil use of Pu
Timeliness	Start	Time to start
	Complete	Time to complete
Operational Effectiveness		
Cost		Investment
		NPV of costs
Environment, Safety & Health		
Human Health & Safety		Public rad. deaths
		Public chem. deaths
		Worker rad. deaths
		Worker chem. deaths
		Transport. deaths
		Accident risk
Protect Natural Environment		Impact on species
		Secondary waste
Economic disruption		New long-term jobs

Technical experts from the national laboratories provided the weights for these specific measures. The data collection process was iterative, giving experts from across the country the opportunity to refine their estimates of these measures. Three committees of experts were formed, one for each of the major categories of alternatives (vitrification, reactors, and boreholes). Some of these measures consisted of even more basic objective measures than given in Table 3. The technical experts assessed the impact of each alternative in their category on each of the measures

Measures and Utilities

In phase II, additional reactor options were generated in an attempt to develop better alternatives. There were a total of nine reactor options at one time in the analysis, as follows:

R1   Existing light water reactor, existing MOX fabrication facilities
R3   Existing light water reactor, new MOX fabrication facilities
       Existing light water reactor, European MOX fabrication facilities
R2   Partially completed light water reactors operated by DOE
R5   CANDU reactors
       CANFLEX
       ABB privately operated MOX fuel reactor
       West privately operated MOX fuel reactor
R4   ABB government operated MOX fuel reactor

The European MOX facilities option was deleted as impractical from a control standpoint. Canadian options were focused on alternative R5. The ABB version of evolutionary light water reactors operated by the Department of Energy was selected to represent that class of reactor.

The development of cost and time estimates for each alternative early in the project is shown in Table 4. The original National Academy of Sciences study expected these life cycle costs to be near 1 billion dollars. The borehole options in fact were in this range, and vitrification options slightly higher. Reactor options were considerably greater in discounted life cycle costs, with the exception of alternative R2, which involved expected revenue from the sale of generated power.

Table 4: Early Cost and Time Estimates for Each Alternative
(estimates changed with design adjustments)

	Investment ($ million)	Life Cycle Cost ($millionNPV)	Time to Start (Year)	Complete (Years)
DIRECT DISPOSAL				
B1direct	777	989	2010	10
B2 immobilized	1124	1084	2011	10
VITRIFICATION				
I1 vitrification-new	1540	2013	2011	9
I2 vitrification-can	840	1067	2005	10
I3 vitrification-adjunct	930	1331	2011	9
I4 ceramic-new	1360	1748	2011	9
I5 ceramic-can	1270	1686	2011	9
I6 electromagnetic	890	1295	2009	10
REACTORS				
R1 existing LWR	1034	3056	2007	22
R2 partially completed	2106	306	2008	19
R3 LWR, new MOX	1585	1967	2008	22
R4 evolutionary	447	5655	2008	18
R5 CANDU	390	2616	2007	25

These estimates were changed over time, as design adjustments were made. These changes were often motivated by adversarial discussions between teams representing each of the three primary classes of alternatives.

These measures were each converted into utilities. The utilities for detailed nonproliferation measures were developed by technical experts, who then conducted multiattribute tradeoff analysis to combine these single-attribute utility functions into utility values for each of the nonproliferation attributes. These tradeoffs were sometimes nonlinear. Utility results for four functions in Table 4 were generated early in the project, and were subsequently adjusted. Single-attribute utility functions for start time and completion time were both linear. For start time (Figure 1) utility was expressed by the line connecting the best measure (year 2007 assigned a utility of 1.0) with the worst measure (year 2011 assigned a utility of 0.0). For completion time (Figure 2), the best measure of 9 years was assigned a utility of 1.0, and the worst measure of 25 years was assigned a utility of 0.0. The function Utility = (25 − measure)/16 gave utility values for the other times of completion.

Figure 1: Utilities for Start Time (Year of Start)

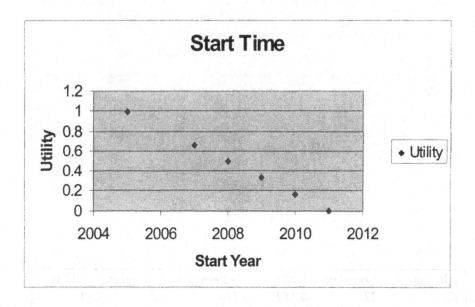

Alternative	Start Time	Utility	Alternative	Start Time	Utility
R1	2007	0.6667	I1	2011	0
R3	2008	0.5	I2	2005	1
Eurofab	2007	0.6667	I3	2011	0
R2	2008	0.5	I4	2011	0
R5	2007	0.6667	I5	2011	0
CANFLEX	2007	0.6667	I6	2009	0.3333
ABB MOX	2008	0.5	B1	2010	0.1667
West MOX	2009	0.3333	B2	2011	0
R4	2008	0.5			

Utility = (2011 − Year)/6

Figure 2: Utilities for Time to Complete (Years)

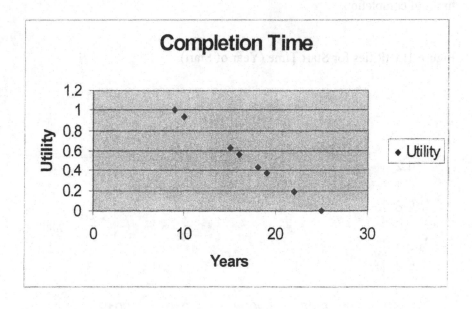

Alternative	Complete	Utility	Alternative	Complete	Utility
R1	22	0.1875	I1	9	1
R3	22	0.1875	I2	10	0.9375
Eurofab	22	0.1875	I3	9	1
R2	19	0.375	I4	9	1
R5	25	0	I5	9	1
CANFLEX	15	0.625	I6	10	0.9375
ABB MOX	18	0.4375	B1	10	0.9375
West MOX	16	0.5625	B2	10	0.9375
R4	18	0.4375			

Utility = (25 – Years)/16

Operational effectiveness measures consisted of a single linear function for Life Cycle Costs (Figure 3), and two linear functions for investment

(Figure 4). For Investment measures below $1000 million, the function was U = (1700 − measure)/1400, a much steeper slope than for measures at or above $1000 million (U = (7000 − measure)/12000).

Figure 3: Utilities for Discounted Life Cycle Costs

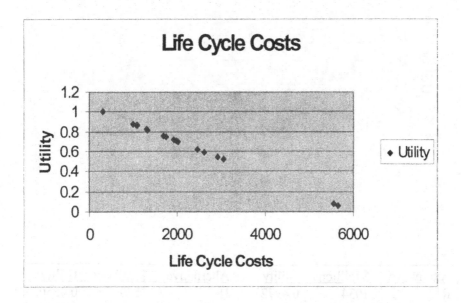

Alternative	$ millions	Utility	Alternative	$ millions	Utility
R1	3056	0.5165	I1	2013	0.6995
R3	1967	0.7075	I2	1067	0.8654
Eurofab	2462	0.6207	I3	1331	0.8191
R2	306	0.9989	I4	1748	0.7460
R5	2616	0.5937	I5	1686	0.7568
CANFLEX	2922	0.54	I6	1295	0.8254
ABB MOX	5655	0.0605	B1	989	0.8791
West MOX	5566	0.0761	B2	1084	0.8625
R4	1924	0.7151			

Utility = (6000 − Life Cycle Cost)/5700 in millions

7

Figure 4: Utilities for Investment Costs

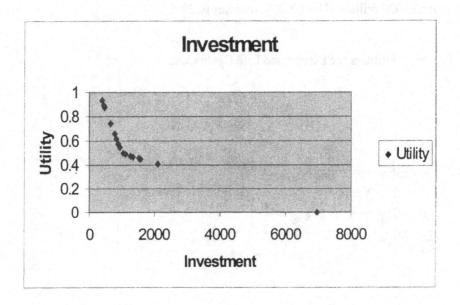

Alternative	$ millions	Utility	Alternative	$ millions	Utility
R1	1034	0.4972	I1	1540	0.4550
R3	1585	0.4513	I2	840	0.6143
Eurofab	1034	0.4972	I3	930	0.5500
R2	2106	0.4078	I4	1360	0.4700
R5	309	0.9357	I5	1270	0.4775
CANFLEX	662	0.7414	I6	890	0.5786
ABB MOX	447	0.8950	B1	777	0.6593
West MOX	472	0.8771	B2	1124	0.4897
R4	6961	0.0033			

If Investment < $1000 million, Utility = (1700 – Investment)/1400 in
        millions

If Investment ≥ $1000, Utility = (7000 – Investment)/12000 in millions

These utility measures are given to demonstrate the methods used. These utility functions evolved over the course of the study, with final utility values given in Table 5.

Table 5: Standardized Scores of Each Alternative on Each Measure

	B1	B2	I1	I2	I3	I4	I5	I6	R1	R2	R3	R4	R5
NONPROLIFERATION													
DOE attractiveness	.00	.00	.00	.00	.00	.00	.00	.00	.00	.00	.00	.00	.00
Other separated materials	.00	.00	.00	.00	.00	.00	.00	.00	.00	.00	.00	.00	.00
Bulk throughput-theft	.06	.06	.06	.06	.06	.06	.06	.09	.06	.06	.06	.06	.06
Num. of processing steps-theft	.85	.71	.65	.65	.65	.65	.59	.59	.38	.38	.38	.38	.38
Max Pu inventory-theft	.96	.96	.90	.90	.96	.90	.96	.96	.99	.99	.99	.99	.99
Number of trips-theft	.55	.55	.55	.55	.55	.55	.55	.55	.05	.05	.10	.05	.05
SST transport miles-theft	.65	.65	.50	.50	.50	.50	.50	.83	.47	.53	.61	.56	.47
Measure uncertainty-theft	.76	.89	.89	.89	.89	.89	.89	.88	.89	.89	.89	.89	.89
Type of NAS-theft	.04	.04	.04	.04	.04	.04	.04	.04	.04	.04	.04	.04	.04
Accessibility of Pu-theft	.00	.00	.00	.00	.00	.00	.00	.00	.00	.00	.00	.00	.00
US Classification	.00	.00	.00	.00	.00	.00	.00	.00	.00	.00	.00	.00	.00
IAEA attractiveness	.00	.00	.00	.00	.00	.00	.00	.00	.00	.00	.00	.00	.00
Bulk throughput-diversion	.06	.06	.06	.06	.06	.06	.06	.09	.06	.06	.06	.06	.06
Processing steps-diversion	.85	.71	.65	.65	.65	.65	.59	.59	.38	.38	.38	.38	.38
Max Pu inventory-diversion	.96	.96	.90	.90	.90	.90	1.0	.50	.99	.99	.99	.99	.99
Measure uncertainty-diversion	.53	.19	.17	.17	.17	.18	.17	.18	.77	.77	.77	.77	.77
Type of NAS-diversion	.04	.04	.12	.12	.12	.12	.12	.12	.04	.04	.04	.04	.04
Accessibility of Pu-diversion	.00	.00	.00	.00	.00	.00	.00	.00	.00	.00	.00	.00	.00
Int'l classification	.00	1.0	1.0	1.0	1.0	1.0	1.0	1.0	.00	.00	.00	.00	.00
Material form	.20	.70	.70	.70	.70	.70	.70	.70	.70	.70	.70	.70	.70
Material location	1.0	1.0	.70	.70	.70	.70	.70	.70	.70	.70	.70	.70	.70
Russian cooperation	.37	.37	.37	.37	.37	.37	.37	.37	.50	.50	.50	.50	.50
Civil use of plutonium	1.0	1.0	.95	.95	.95	.95	.95	.95	.70	.70	.70	.70	.10
Time to start	.67	.67	.53	.87	.53	.53	.87	.47	.73	.47	.47	.40	.67
Time to complete disposition	.67	.67	.73	.60	.73	.73	.60	.73	.33	.33	.13	.40	.40
OPERATIONAL EFFECTIVENESS													
Investment costs	.84	.81	.71	.92	.85	.74	.92	.83	.86	.56	.80	.02	.88
Net present value of costs	.63	.49	.36	.74	.54	.42	.74	.57	.70	.70	.69	.09	.59
ENVIRONMENT, SAFETY & HEALTH													
Public radiation deaths	.50	.50	.50	.50	.50	.50	.50	.49	.68	.31	.68	.31	.50
Public chemical deaths	1.0	1.0	1.0	1.0	1.0	1.0	1.0	1.0	1.0	1.0	1.0	1.0	1.0
Worker radiation deaths	.32	.30	.34	.34	.34	.34	.34	.32	1.0	.00	1.0	.00	.34
Worker chemical deaths	1.0	1.0	1.0	1.0	1.0	1.0	1.0	1.0	1.0	1.0	1.0	1.0	1.0
Transportation deaths	.55	.28	.61	.61	.61	.60	.60	.59	.56	.56	.56	.56	.77
Accident risk	1.0	1.0	1.0	1.0	1.0	1.0	1.0	1.0	.98	.98	.98	1.0	1.0
Impact on species	.88	.88	.88	.88	.88	.88	.88	.88	.88	.88	.88	.88	.88
Secondary waste generation	.54	.53	.54	.54	.54	.54	.54	.53	.46	.14	.46	.19	.50
Economic disruption	1.0	1.0	1.0	1.0	1.0	1.0	1.0	1.0	1.0	.88	1.0	.79	1.0
New long-term jobs	.63	.61	.77	.77	.77	.80	.80	.54	.60	.87	.60	.87	.60

For instance, the utility for investment of newly construction evolutionary light water reactors (alternative R4) initially was quite high, based on the assumption that it would be privately built. The final utility for investment in this alternative was very low, reflecting use of government funds for construction.

Multiple Attributes, Multiple Parties

This problem consisted of three major attributes (nonproliferation-security, operational effectiveness, and environment-safety-health). Each of these attributes was subdivided, as discussed below. There were a number of parties involved in the decision. The Department of Energy had the primary decision. The National Academy of Sciences had voiced concern about the decision, representing the public at large. National laboratories had a part in the decision, both as sources of technical expertise, as well as proponents for various types of options. The Russian Government was involved in the decision, because as treaty participant, they have a parallel decision to make, and also had an interest in the U.S. decision.

Parties and Interests in the Plutonium Disposition Decision			
Criteria	Department of Energy	National Academy of Sciences National Labs	Russian Government
Nonproliferation	x	x	x
Security	x		x
Time	x	x	x
Cost	x	x	
Envir., Health & Safety	x	x	

The sequence of major events can be displayed in rough form.

Sequence of Major Events: Plutonium Disposition

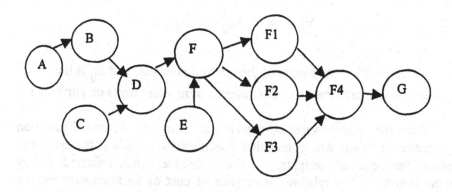

A   Strategic Arms Limitation Treaty creates need to dispose of
             warheads, 1993
B   Generation of initial set of 36 alternatives
C   National Academy of Sciences calls for building reactor,
             vitrification during analysis
D   Original list of 36 alternatives screened to 11, 1994
E   Laboratories generate additional alternatives, total up to 17, 1995
F   Multiattribute analysis, 1995-1996
F1   Technical study of reactor options by Reactor Team
F2   Technical study of vitrification options by Vitrification Team
F3   Technical study of borehole options by Borehole Team
F4   MAUT reconciliation
...   iterations with improved decision alternatives
G   Department of Energy selects existing reactor and vitrification
             options, December 1996

Multiattribute Utility Analysis

Multiattribute analysis aggregates the single-attribute utility values and their
interactions across dimensions through a weighting procedure. The
simplest and most widely used aggregation rule is to take a weighted linear
average. This form was appropriate here because public decisions
involving positions of different groups of people have been noted to be
linear, and it has been argued that they should be linear.

The model:

$$\text{Value} = U_j = \sum_{i=1}^{k} W_i S_{ij}$$

where W_i is the weight of the kth of n attributes and S_{ij} is the measure of the performance of alternative j of m alternatives on attribute i.

When the criteria were more technical in nature, i.e. the disposition safeguards of an alternative, technical specialists were used as the experts to assess the required weights.   For objectives that reflected policy implications, i.e. the relative importance of cost of an alternative vs. the alternative's impact on the environment, OFMD personnel provided the tradeoffs necessary to generate the weights W_i used to aggregate the individual criteria scores. The exact magnitude of the assessed weights may be legitimately questioned (see for example, Weber and Borcherding, 1993). Therefore, sensitivity analysis was applied to determine the impact of relative weights provided to the three criteria at the top of the hierarchy: nonproliferation, operational effectiveness, and environment safety & health. OFMD management specified that the weight given to environment safety & health should not exceed either of the other two criteria. Sensitivity analysis was used to explore all of the other possible combinations of relative weights.

Conducting a sensitivity analysis of the weights is often insightful, but standard techniques typically vary a single weight, holding the ratio among other weights constant, and observe the effect on the results of the model. Such a one-dimensional approach can miss certain relationships among decision criteria. A method of simultaneously varying all, or at least a large subset, of the weights was obtained by applying Monte Carlo simulation (see Butler, Jia, and Dyer, 1997).

Monte Carlo Simulation

Simulation provides a flexible means to test the sensitivity of the weights of a decision problem. Butler, Jia and Dyer (1997) outline three classes of simulation models that offer assistance in the evaluation of weights for multi-criteria models, including the possibility of a multiplicative aggregation model. The three classes are random weights (no importance information), order weights (rank order of attribute importance), and treating the assessed weights as responses from a possible distribution. The second class of random weights was used, as OFMD specified constraints on the weight given to environment, safety and health. The rank order of weights on the measures was maintained, but the weights were otherwise generated at random. The results of a 5,000 iteration simulation of the attribute weights are given in Table 6.

Table 6: Simulation Results

Alternative	Utility rank	Best rank	Worst rank	Mean Rank	Mode rank	95% Perc
I5 Ceramic Can-in-Can	1	1	3	1.19	1	2
I2 Vitrif. Can-in-Can	2	1	4	2.19	2	3
R1 Exist LWR, Exist Fac	3	1	8	2.83	3	4
R5 CANDU	4	3	13	6.73	7	11
B1 Borehole (Direct)	5	3	8	4.97	5	6
R3 Exist LWR Green Fac	6	2	12	6.16	4	10
B2 Borehole (Immob)	7	1	9	6.95	9	9
I3 Vitrif. Adj. Melter	8	3	8	5.98	6	7
I6 Electrometallurgical	9	7	12	8.51	8	10
I4 Ceramic Greenfield	10	6	11	9.74	10	10
R2 Part Comp LWR	11	9	12	11.84	12	12
I1 Vitrif. Greenfield	12	7	12	10.91	11	12
R4 Evolutionary LWR	13	11	13	13.00	13	13

Reactor Options:

Reactor options involve preparing the plutonium to a form usable in the reactor for generation of power. Not all plutonium is burned (in fact, only about 7 percent), and secondary waste is generated which must be disposed of in a storage facility. The relative advantages and disadvantages by reactor alternative are:

	Relative advantage	Relative disadvantage
R1 Exist LWR, Exist Fac	Start time	SST transport miles
	Investment	
	Public radiation deaths	
	Worker radiation deaths	
	Boom/bust	
R2 Part Comp LWR	Life-cycle costs	Public radiation deaths
	New long-term jobs	Worker radiation deaths
		Secondary waste gen.
R3 Exist LWR, Green Fac	Number of trips	Time to complete
	SST transport miles	
	Public radiation deaths	
	Worker radiation deaths	
	Economic disruption	
R4 Evolutionary LWR	Time to complete	Maximum Pu inventory
	Accident risk	Time to start
	New long-term jobs	Investment cost
		Life-cycle costs
		Public radiation deaths
		Worker radiation deaths
		Secondary waste gen.
		Boom/bust
R5 CANDU	Time to complete	SST tranportation miles
	Investment cost	Civil use of PU
	Transport deaths	
	Accident risk	
	Secondary waste generated	
	Boom/bust	

Reactor alternative R1, an existing light water reactor with existing preparation facilities, ranked the highest among reactor alternatives for all weight combinations. Alternative R1 has strong advantages in start time and investment costs, obvious from the fact that new construction would be minimal.

Reactor alternative R3, an existing light water reactor with greenfield preparation facilities, was the second most attractive reactor alternative for most sets of weights, the exception being the weight set where all weight was given to nonproliferation issues. Alternative R5, the Canadian reactor option, had a relatively high utility analysis ranking, but did not do as well in any of the weight combinations used. Alternative R4, an advanced reactor requiring construction, ranked last of all 13 options for all weight sets except when all weight was given to nonproliferation issues.

Vitrification Options:

Vitrification involves encasing plutonium, suspending it in such a way that it is not heavily concentrated. The vitrified material must then be disposed of. The relative strengths and weaknesses of these options is given below:

	Relative advantage	Relative disadvantage
I1 Vitrification Greenfield		Life-cycle cost
I2 Vitrificat. Can-in-Can	Time to start	Time to complete
	Investment costs	
	Life-cycle costs	
I3 Vitrificat. Adj. Melter	Maximum Pu inventory	Time to start
I4 Ceramic Greenfield	Measurement uncertainty	Investment
	New long-term jobs	Life-cycle cost
I5 Ceramic Can-in-Can	Maximum Pu inventory	No. of processing steps
	Time to start	Time to complete
	Investment	
	Life-cycle costs	
	New long-term jobs	
I6 Electrometallurgical	Bulk throughput	No. of processing steps
	SST transport miles	Time to start
		Public radiation deaths
		Worker radiation deaths
		Transportation deaths
		Secondary waste generat.
		New long-term jobs

The ceramic can-in-can option I5 was best among immobilization options under almost all weight sets tested. The vitrification can-in-can option I2 was almost always second. The other options were always inferior, due in great part to cost advantages of the ceramic and vitrification can-in-can options.

Borehole Options:

Between the two borehole options, the immobilization option B2 was stronger on safety and security measures, while the direct borehole option B1 had relative advantage with respect to cost items and environmental safety and health measures.

	Relative advantage
B1 Borehole (Direct)	investment costs
	life-cycle costs
	number of processing steps
B2 Borehole (Immobilized)	international classification
	material form
	worker radiation deaths
	transportation deaths
	secondary waste generated
	new long term jobs

The borehole immobilization option B2 ranked higher than the direct borehole option B1 when the weights on safety and security measures were higher. The direct borehole option benefited from having lower costs in other weight combinations.

Insights from Sensitivity Analysis Methods:

The vitrification options compared quite favorably with both reactor and borehole options. They were second to reactor option R1 when the cost criteria were given lower weights. When high weight was given to safety and security criteria, the borehole option with immobilization (B2) ranked higher.

The DOE record of Decision (DOE-ROD 97) recommended that a hybrid approach be used to dispose of surplus weapon's grade plutonium. This hybrid approach would consist of an existing reactor (R1 - using existing facilities) and either vitrification (I2) or ceramic immobilization (I5) at an existing facility. The CANDU alternative (R5) was viewed a back-up alternative should negotiations with Russia lead to joint disposition activities in Canada. Thus, DOE's recommendation is consistent with the results of the sensitivity analysis on the weights. The decision to adopt the original National Academy of Sciences recommendation of a vitrification alternative and a reactor alternative simultaneously was given by the outgoing Energy Secretary shortly after the 1996 U.S. elections.

Conclusions

The multi-attribute analysis clearly and systematically identified factors of importance, obtained scientifically valid measures, converted these measures to utilities using qualified people, and tested the impact of different policy weights. The simulation approach to testing the sensitivity of selection problems provided insight into the decision problem and supported DOE's final recommendation. Negative conclusions are that the original investment estimate of $1 billion is now about $2 billion, worse than all but one of the thirteen options analyzed. Life cycle cost can't be added in this case, because the work is shared by the two methods adopted, but the cost of each option cannot be halved either. As alternative R1 ranked 12th on cost originally, cost was obviously not the driver of the decision.

The project to analyze options (Phase II) was originally scheduled to take less than a year. There was considerable slippage, and the analysis took almost two years. There was also considerable change in estimates as time passed, as noted in the earlier comments about cost and time estimates. Such estimates are expected to inflate, and the figures used in Phase II will undoubtedly be much higher before all of the plutonium has been processed and stored. Drezner (1993) reported consistent project cost overruns for 197 U.S. Air Force projects as of December 1990. Merrow (1988) found an average cost growth of 88 percent in large projects, and average cost growth of 157 percent for nuclear projects. Given the extremely long duration of this project, and the fact that not one but two options were adopted, the cost overruns should be spectacular.

The analysis identified the conditions under which each of the three disposition options appeared the most favorable. The borehole option was the most attractive only if all of the weight were given to nonproliferation factors. If large weight was given to operational effectiveness, the reactor options ranked higher. If large weight was given to environment, safety and health issues (this weight was limited to not receive greater weight than either nonproliferation or operational effectiveness), the reactor options again ranked relatively higher. In twelve of the twenty six weight combinations the vitrification options were ranked highest. These involved having larger weights given to nonproliferation factors.

References

Butler, J, Jia, J. and Dyer, J. Simulation Techniques for the Sensitivity Analysis of Multi-criteria Decision Models. *European Journal of Operational Research*, vol. 103, no. 3, 1997, 531-546.

DOE-ROD 97 *Record of Decision for the Storage and Disposition of Weapons-Usable Fissile Materials Final Programmatic Environmental Impact Statement.* United States Department of Energy; Office of Fissile Materials Disposition, January 14, 1997.

DOE-SCR 95 *Summary Report of the Screening Process to Determine Reasonable Alternatives for Storage and Disposition of Weapons-usable Fissile Materials, DOE/MD-0002.* United States Department of Energy; Office of Fissile Materials Disposition, March 17, 1995.

Drezner, J.A. *An Analysis of Weapon System Cost Growth.* Rand Corporation, Santa Monica, CA, 1993.

Dyer, J., Edmunds, T., Butler, J. C., and Jia J. Evaluation of Alternatives for the Disposition of Surplus Weapons-usable Plutonium. Amarillo National Resource Center for Plutonium Technical Paper 1997-1. http://www.pu.org/main/reports/reports.html

Dyer, J., Edmunds, T., Butler, J.C., and Jia, J. A multiattribute utility analysis of alternatives for the disposition of surplus weapons-grade plutonium. *Operations Research* vol. 46, no. 6, 1998, 749-762.

Garwin, R.L. Options and Choices in the Disposition of Excess Weapon Plutonium. *Materials Research Society Symposium Proceedings*, vol. 412, 1996, 3-13.

Holdren, J. P., Ahearne, J. F., Budnitz, R. J., Garwin, R. L., May, M. M., Pigford, T. H. AND Taylor, J. J. *Management and Disposition of Excess Weapons Plutonium: Reactor-Related Options*, National Academy of Sciences, Washington, D. C., 1995.

Merrow, E.W. *Understanding the Outcome of Megaprojects: A Quantitative Analysis of Very Large Civilian Projects.* Rand Corporation, Santa Monica, CA, 1988.

OFMD. Summary Report of the Screening Process to Determine Reasonable Alternatives for Storage and Disposition of Weapons-Usable Fissile Materials. Final Draft dated March 17, 1995, Washington, D. C.

Weber, M. and Borcherding, K. Behavioral Influences on Weight Judgments in Multiattribute Decision Making. *European Journal of Operational Research*, **67**, 1993, 1-11.

5 PROJECT SELECTION AND CONTROL

We have seen a number of instances of selection of an alternative from a small number of alternatives. In this chapter we will look at two applications of the analytic hierarchy process (AHP) to this type of decision problem. The first application used AHP in a control problem for a pipeline project (Dey, et al., 1994, 1996, 1998). We will discuss different multicriteria pipeline applications further in Chapter 7. The second AHP application involved water planning and development projects (Srinivasa Raju and Pillai, 1999a, 1999b). AHP has been applied to many decision problems in business, as discussed in Chapter 2. It has been applied to other public works policy decisions as well, notably Hämäläinen (1988, 1990, 1991) who developed a decision support system for Finnish leaders to study tradeoffs in electrical generation alternatives.

Petroleum Pipeline Project Analysis

Many large-scale projects involve high levels of uncertainty, requiring thorough risk analysis during the planning stage. Dey, et al. (1994) used AHP to analyze and assess project risks combining both objective and subjective data in a pipeline location decision in India.

Construction projects include many uncertainties, including aspects of design, materials, equipment availability, contractor ability, climate, economics, and politics. Uncertainties tend to increase with project size, project complexity, and the number of external agencies involved. Additional uncertainties arise from environmental issues and currency fluctuations. A cross-country petroleum pipeline construction project faces all of these factors. Conventional cost-benefit analysis of such a project is inadequate, because the project management team is not able to gather all of the needed risk data early enough to accurately estimate risk implications. Dey, et al. (1994) identified risk factors and used AHP instead of probability analysis, Monte Carlo simulation, or fuzzy set analysis because of the lack of detailed quantitative information required by those methods, and because of the need to include subjective data.

Multiple Attributes, Multiple Parties

The Dey, et al. (1994) analysis was a proposed approach, demonstrated with a pipeline project. The multiple actors and their interests given here are therefore generic. The criteria were provided by Dey, et al.

Parties and Criteria			
Criteria	Pipeline Management	Contractor	Government
Technical risk	x	x	
Financial, economic & political	x		
Disaster	x	x	x
Legal requirements			x

The first step was to develop a hierarchy for the decision problem. This was based on the project work-breakdown structure, applying risk analysis for each work package. Risk input was obtained through interviews with managers. Relative risk was expressed as high, medium, or low. Likelihoods of risk levels were estimated based on pairwise comparisons. A sensitivity analysis was conducted for relative judgments at various elements of the hierarchy. AHP yielded an overall risk estimate for each work package, which was combined with estimates of the likelihood levels for each of these risks.

Hierarchy

The overall hierarchy of riskiness of pipeline construction was divided into four primary risk areas, each of which was further broken down to more specific risks. Using management expertise, relative risk ratings were for each of the four hierarchical levels. There were three levels of project risk, with the top level being overall project risk, a second level of four general risk areas, and a third level of thirteen more detailed risk elements:

Overall project risk	
Technical risk	
Design change	0.397
Equipment	0.125
Materials	0.079
Financial, economic and political risk	
Inflation	0.005
Subcontractor failure	0.007
Unavailability of funds	0.011
Local laws	0.029
Acts of God	
Earthquake	0.019
Rain and flood	0.062
Subsidence and collapse 0.017	
Statutory Requirements	
Environmental clearance0.019	
Right-of-way acquisition0.083	
Permits for crossings	0.147

Below each of these thirteen third-level factors were the three risk levels of High, Medium, and Low. Overall risk for each work package was determined by its own set of pairwise comparisons of the three risk levels. Aggregation for each work package provided an overall risk rating for that work package. Contingency allowances were allocated using a model reflecting not only the AHP input, but also PERT analysis of the most likely, optimistic, and pessimistic .cost estimates for project activities. Contingency funds were allocated to match the percentage of success required by project management.

The sequence of major events would generally consist of the following activities:

A   Project adoption
B   Development of work breakdown structure
C   Risk analysis
D   Contingency appropriation
E   Construction

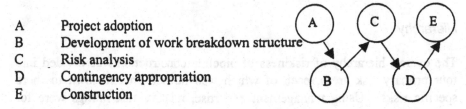

The system provided project control in that if a work package turned out to require additional funds beyond the contingency allowance for the work package, the project manager documented the reasons for the change. Top management then either reduced the scope of the project, or arranged for additional funding. If a particular work activity was completed under budget, excess budget was transferred back into the contingency reserve.

The AHP portion of the analysis focused management attention on the most likely sources of risk, providing insight into the project's scope of work. AHP was seen as providing a comprehensive framework for project management to think through the decision problem. It was this systematic thinking environment that was viewed as the greatest contribution of AHP, leading to greater confidence in project achievement.

Water Resource Planning

Water resources once were so plentiful that water was typically used as an example of the freest of goods. Today, unfortunately, very few parts of the world find water that plentiful. The management of water resources has become a critical area of policy management. There have been a number of studies applying multicriteria decision analysis to water resources planning (Gershon and Duckstein, 1983; Ko, et al., 1994, Srinivasa Raju and Pillai, 1999b). Srinivasa Raju and Pillai (1999a) applied AHP (comparing results with other multicriteria methods) to the problem of selecting the best reservoir configuration for the Chaliyar river basin, Kerala, India.

Chaliyar River Basin Planning

The Chaliyar River flows into the Arabian Sea from the State of Kerala in India. The 170 kilometer long river has nine major tributaries. There are 93,276 hectares of land in the river basin, with irrigation facilities serving only a limited area of paddy lands. Additional cultivation of crops would be possible with more irrigation facilities. Srinivasa Raju and Pillai (1999a) studied the configuration of reservoirs to develop the Chaliyar basin. The study involved six active criteria, including irrigation, production of power, supply of drinking water, environmental quality, flood protection, and project benefits. There were nine reservoir component sites, which could be combined in various ways, yielding 27 alternative configurations in total (see Table 1). Eight configurations were selected for analysis, two of which (alternative 5 and alternative 7) had identical performances on the six criteria used in the analysis. Both alternatives were retained because they had difference performances on the additional criteria of employment generation and the number of people effected by the project. These last two criteria were potentially important, and alternatives 5 and 7 each had a relative advantage on these two criteria.

Table 1: Criteria/Attainment Matrix – Chaliyar River Basin

	IR	PO	DW	EQ	FL	BE
weights	10	8	5	8	6	7
Alternatives						
A1:R1,R3,R4,R5,R7,R8	60	15	Good	Average	40	20
A2:R2,R4,R7,R8,R9	60	10	Good	Average	50	20
A3:R2,R3,R4,R5,R9	60	15	Good	Average	40	30
A4:R3,R5,R7,R9	20	5	Average	Good	30	60
A5:R2,R4,R5,R8	70	15	Average	Average	30	20
A6:R2,R4,R7.R8	60	10	Average	Average	40	30
A7:R4,R5,R8,R9	70	15	Average	Average	30	20
A8:R2,R6,R9	20	5	Average	Good	30	50

Criteria		Reservoir Sites	
IR	irrigation	R1	Irujhupuzha
PO	power production	R2	Iruvanjipuzha
DW	drinking water	R3	Kanhirapuzha
EQ	environmental quality	R4	Chalipuzha
FL	flood protection	R5	Maruthpuzha
BE	project benefits	R6	Punnapuzha
		R7	Karimpuzha
		R8	Kuthirapuzha
		R9	Arecode

Multiple Attributes, Multiple Parties

As with the prior case, the multiple actors and their interests given here are generic. The criteria were provided by Srinivasa Raju and Pillai (1999a).

Parties and Criteria

Criteria	Farmers	Pipeline Businesses	Inhabitants	Gov't
Irrigation	x			
Power production		x	x	
Drinking water			x	
Environment			x	x
Flood protection	x	x	x	x
Economic benefits		x		x

The sequence of project development is similar to that described in the prior case.

A   Evaluation of each site
B   Development of packages of sites
C   Risk analysis
D   Contingency appropriation
E   Construction

Hierarchy

The first step of AHP is to develop the hierarchy of criteria and alternatives, which determines the pairwise comparison matrices required. In the Chaliyar River case, there were six criteria of importance, as displayed in the column headings in Table 1. These criteria were all independent, so they could all be included at one level. Below each of these six criteria each of the combinations of reservoirs generated as alternative solutions were included as hierarchy elements. Srinivasa Raju and Pillai had eight alternatives. For the data in Table 1, a number of dominated solutions are included. (A dominated solution is one that is no better than its dominating solution on any criterion, and is worse on at least one criterion – examples here include the 1st alternative being dominated by the 3rd, the 6th alternative dominated by the 3rd, and the 8th alternative dominated by the 4th. However, as with alternatives 5 and 7, other criteria could make the currently dominated alternatives more attractive than the other seven alternatives. The top level of the hierarchy is used to estimate relative weights for the six criteria in one pairwise comparison matrix, while the bottom six pairwise comparison matrices were used to generate the single-attribute utility functions. Srinivasa Raju and Pillai (1999a) allowed fractional pairwise comparison ratios, reflecting attainments close in value.

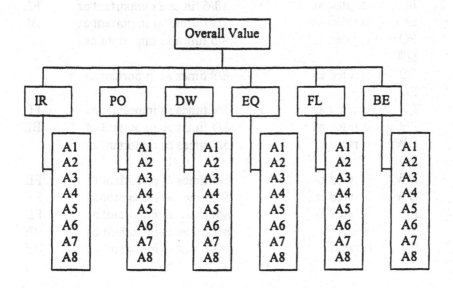

Subjective Pairwise Comparisons

The next phase in the AHP analysis is to obtain the relative importance of factors within each hierarchical level for each factor. This is accomplished through a pairwise comparison of all elements in that section of the hierarchy. One pairwise comparison would be required for overall utility, comparing reimbursement, type of work, and personal factors. Assuming they are ranked in order of greatest preference first, the upper right triangular portion of the following matrix would be completed with the strength of preference given according to the scale in the adjacent table. Even (or fractional) numbers can be used for ratings between the other ratings.

This AHP model requires seven sets of pairwise comparisons. One of these sets is to allocate the proportional weights for the four criteria. Srinivasu Raju and Pillai used the same calculation applied in MAUT (see Chapter 4) to calculate relative weights of these six factors:

IR	is rated as	10/8 times as important as	PO
IR	is rated as	10/5 times as important as	
DW			
IR	is rated as	10/8 times as important as	
EQ			
IR	is rated as	10/6 times as important as	FL
IR	is rated as	10/7 times as important as	BE
PO	is rated as	8/5 times as important as	
DW			
PO	is rated as	8/8 times as important as	
EQ			
PO	is rated as	8/6 times as important as	FL
PO	is rated as	8/7 times as important as	BE
DW	is rated as	5/8 times as important as	
EQ			
DW	is rated as	5/6 times as important as	FL
DW	is rated as	5/7 times as important as	BE
EQ	is rated as	8/6 times as important as	FL
EQ	is rated as	8/7 times as important as	BE
FL	is rated as	6/7 times as important as	BE

The ways to calculate relative scores or weights include simply averaging the three weights given above, which is probably the simplest method, but relatively unstable. Another approach would be obtaining the geometric mean (see next chapter), which has theoretical advantages (Barzilai, et al., 1987), as well as being relatively easy computationally. The eigen vector provides a robust estimator, and is justification argument is provided by Saaty (1977). The eigen vector resulting from the pairwise comparisons of criteria given above is:

IR	0.227
PO	0.182
DW	0.114
EQ	0.182
FL	0.136
BE	0.159

The performance measures for each of the eight criteria were given in Table 1. Developing utility functions would require six pairwise comparison matrices of these seven alternatives.
The Irrigation pairwise comparison matrix:

IR	A2	A3	A4	A5	A6	A7	A8
A1	1	1	3	6/7	1	6/7	3
A2		1	3	6/7	1	6/7	3
A3			3	6/7	1	6/7	3
A4				1/3.5	1/3	1/3.5	1
A5					7/6	1	3.5
A6						6/7	3
A7							3.5

Normalizing these continuous numbers would involve no possible inconsistency, because each row is the same set of relative ratios.
The overall rating obtained is:

A1	0.143
A2	0.143
A3	0.143
A4	0.048
A5	0.167

	A6	0.143
	A7	0.167
	A8	0.048

Similar calculations for all six criteria yielded the following matrix, with was aggregated using the weights of the six criteria derived above.

	IR	PO	DE	EQ	FL	BE	Total
Weight	0.227	0.182	0.114	0.182	0.136	0.159	
A1	0.143	0.167	0.182	0.111	0.138	0.080	0.135
A2	0.143	0.111	0.182	0.111	0.172	0.080	0.130
A3	0.143	0.167	0.182	0.111	0.138	0.120	0.142
A4	0.048	0.056	0.091	0.167	0.103	0.240	0.114
A5	0.167	0.167	0.091	0.111	0.103	0.080	0.126
A6	0.143	0.111	0.091	0.111	0.138	0.120	0.121
A7	0.167	0.167	0.091	0.111	0.103	0.080	0.126
A8	0.048	0.056	0.091	0.167	0.103	0.200	0.107

This yields a rank-order of A3, A1, A2, A5=A7, A6, A4, and A8. AHP analysis therefore would yield a recommendation of reservoir sites R2, R3, R4, R5, and R9. In the Srinivasa Raju and Pillai study, his recommendation matched results obtained from PROMETHEE, compromise programming, and another method. ELECTRE II yielded a different rank order.

Summary

AHP has been very widely applied in the U.S. and Canada for many business related decisions, especially those involving multiple criteria including subjective estimates of tradeoff information. It has not been as widely used in public policy decisions, but has been applied in some cases as demonstrated by the two applications we have presented. As Belton and Gear (1983) and Barzilai, et al. (1987) have pointed out, AHP is subject to rank reversal if additional alternatives are added to the analysis. (Saaty, 1987 provides counterarguments about the appropriateness of rank reversal.) AHP is quite popular in the United States and Canada. A number of improvements to the method have been proposed by Barzilai (1998), Lootsma (1993), Lootsma and Schuijt (1997), Bana e Costa and Vansnick (1997), and others.

References

Barzilai, J. On the decomposition of value functions, *Operations Research Letters* **22**, 1998, 159-170.

Barzilai, J., Cook, W.D. and Golany, B. Consistent weights for judgements matrices of the relative importance of alternatives, *Operations Research Letters* **6**:3, 1987, 131-134.

Bana e Costa, C.A. and Vansnick, J.C. Applications of the MACBETH aapproach in the framework of an additive aggregation model, *Journal of Multi-Criteria Decision Analysis* **6**:2, 1997, 107-114.

Belton, V. & Gear, T., On a shortcoming of Saaty's method of analytical hierarchies, Omega 11:3, 1983, 227-230.

Dey, P.K., Tabucanon, M.T., and Ogunlana, S.O. Planning for project control through risk analysis: A petroleum pipeline, *International Journal of Project Management* **12**:1, 1994, 23-33.

Dey, P.K., Tabucanon, M.T. and Ogunlana, S.O. Petroleum pipeline construction planning: A conceptual framework, *International Journal of Project Management* **14**:4, 1996, 231-240.

Dey, P.K., Ogunlana, S.O., Gupta, S.S. and Tabucanon, M.T. A risk-based maintenance model for cross-country pipelines. *Cost Engineering* **40**:4, 1998, 24-31.

Gershon, M., and Duckstein, L. Multi objective approaches to river basin planning. *Journal of Water Resources Planning management* **109**:1, 1983, 13-28.

Hämäläinen, R.P. Computer assisted energy policy analysis in the Parliament of Finland, *Interfaces* 18:4, 1988, 12-23.

Hämäläinen, R.P. A decision aid in the public debate on nuclear power, *European Journal of Operational Research* **48**:1, 1990, 66-76.

Hämäläinen, R.P. Facts or values – how do parliamentarians and experts see nuclear power? *Energy Policy* **19**:5, 1991, 464-472.

Ko, S.K., Fontane, D.G. and Margeta, J. Multiple reservoir system operational planning using multi-criterion decision analysis. *European Journal of Operational Research* **76**, 1994, 428-439.

Lootsma, F.A. Scale sensitivity in a multiplicative variant of the AHP and SMART, *Journal of Multi-Criteria Decision Analysis* 2:2, 1993, 87-110.

Lootsma, F.A. and Schuijt, H. The multiplicative AHP, SMART, and ELECTRE in a common context, *Journal of Multi-Criteria Decision Analysis* 6:4, 185-196

Saaty, T. L., Rank generation, preservation, and reversal in the Analytic Hierarchy Process. *Decision Sciences* 18, 2, 1987, 157-177.

Srinivasa Raju, K. and Pillai, C.R.S. Multicriterion decision making in river basin planning and development, *European Journal of Operational Research* 112, 1999a, 249-257.

Srinivasa Raju, K. and Pillai, C.R.S. Multicriterion decision making in performance evaluation of an irrigation system, *European Journal of Operational Research* 112, 1999b, 479-488.

6 SOLID WASTE MANAGEMENT SYSTEM SELECTION

Researchers Jonas Hokkanen and Pekka Salminen have applied outranking ELECTRE methods to a series of solid waste management systems in Finland. Their work demonstrates some of the factors involved in this ecologically important decision and some of the issues involved in public decision making for these hard issues. They also demonstrate the use of a variety of versions of the ELECTRE method, and argue the conditions under which each outranking model is appropriate.

There are at least six versions of the ELECTRE family (Roy and Bouyssou, 1993), each unique, but all based on similar general ideas. Outranking relations are constructed representing the decision-maker's strongly established preferences based on the available information. Various mathematical functions are available to indicate the degree of dominance of one alternative over another on each criterion. This makes it possible to analyze both quantitative and qualitative data. The method allows individual pairs of options to remain uncompared where there are very conflicting criteria estimates (Rogers and Bruen, 1998a). Initial weights are used within some of the ELECTRE methods (not ELECTRE IV), but these weights are meant to reflect relative importance of criteria rather than constants of scale (Vincke, 1992; Rogers and Bruen, 1998b).

Uusimaa Region Study

The first of the applications reported involved selection of a municipal solid waste management system for the province of Uusimaa in southern Finland (Hokkanen, et al., 108). This system was to serve 38 municipalities including about 1.23 million people, two-thirds of whom lived in Helsinki. This system needed to process about 620,000 tons of solid waste (presumably per year). Alternative waste treatment methods considered were sanitary landfill, incineration, fermentation, and composting. Each method would comply with anticipated Finnish environmental legislation.

A second aspect of each alternative was the degree of centralization. Decentralized systems would have a facility serving each municipality. A centralized administration would involve one facility. In between these two options was an intermediate arrangement, with each province divided into cooperation areas. Since some of the methods required higher volumes to be viable, not all pairs of the four methods and three administrative arrangements were used. The alternatives considered were as given in Table 1.

Table 1: Solid Waste Disposal Alternatives for the Uusimaa Region

Treatment Method	Cooperation Level	Sites
I Incineration	Centralized	1
II RDF combustion	Intermediate	4
III Fermentation	Intermediate	4
IV Biothermic (RDF + fermentation)	Intermediate	4
V Composting plant	Intermediate	4
VI Landfill	Intermediate	4
VII Landfill	Decentralized	32
VIII Landfill and open composting	Decentralized	32

Criteria Development

A supervisory group consisting of twelve people was formed to determine criteria for waste management system evaluation. This group included representatives of municipalities, municipal councils, regional planning associations and water and environment districts. Decision making authority rested with municipal councils. The supervisory group selected the criteria given in Table 2.

Table 2: Uusimaa Region Solid Waste Disposal Criteria

g1 Political feasibility	public acceptance
g2 Technical reliability	feasibility, maturity, suitability for local conditions
g3 Transport reliability	collection and transport feasibility
g4 Benefits to national economy	fossil fuels saved
g5 Employment	number of employments
g6 Acute environmental effects	heavy metal releases to the air
g7 Long-term environmental effects	greenhouse effects, acidic and surface water releases
g8 Environmental hygiene	dust, noise
g9 Resource recovery level	recycled and incinerated waste
g10 Costs	total annual cost/ton of waste

Weights

A questionnaire was sent to the manager of environmental affairs and the manager of technical affairs of each municipality. Of the 70 questionnaires sent out, 53 were returned. Respondents were asked to assign weights between 1 and 10 to each of the criteria. Results are recapped in Table 3. Hokkanen, et al. stated that while respondents understood the criteria, the weights they gave did not necessarily exactly reflect criteria importance. The respondents did not have experience in judging the importance of criteria in the numerical form asked for.

Table 3: Recapitulation of Weight Questionnaire Responses – Uusimaa
Region

		Mean	Std Dev	Min	Max
g1	Political feasibility	4.8	2.8	1	10
g2	Technical reliability	7.7	2.0	2	10
g3	Transport reliability	7.1	2.0	2	10
g4	Benefits to national economy	5.9	2.3	1	10
g5	Employment	3.4	2.2	1	9
g6	Acute environmental effects	9.1	1.1	5	10
g7	Long-term environmental effects	9.4	1.0	6	10
g8	Environmental hygiene	8.3	1.5	2	10
g9	Resource recovery level	7.6	1.9	3	10
g10	Costs	7.3	1.9	3	10

The weights obtained from the 53 respondents were highly variable,
reflecting wide differences in opinion. There was less disagreement about
the more important measures, such as impact on the long-term environment
and hygiene.

Measures

A questionnaire was mailed to 21 Finnish experts in waste management,
and 17 responses were received. Each expert estimated the degree, on a 1-
10 scale, how well each of the alternatives satisfied each of the first eight
criteria. Responses are recapped in Table 4, which gives both means and
lower quartile values. The lower quartile responses were used in the
multicriteria analysis, assuring that 75 percent of the experts considered that
the alternative had a performance at least as good as the rating used for each
criterion. Resource recovery level and cost were computed for each
alternative. Resource recovery level was the estimated percentage of the
difference between total waste volume and landfill volume to total waste
volume. Cost was estimated per ton of waste, and was adjusted for income
from resource recovery.

Table 4: Project Impact Matrix – Uusimaa Region

Alternative	g1	g2	g3	g4	g5	g6	g7	g8
				Means				
I Incineration	2.5	8.2	7.7	6.7	3.2	7.2	5.1	7.2
II RDF combustion	4.8	6.3	7.6	6.0	4.9	6.4	6.9	6.4
III Fermentation	4.5	3.1	6.4	4.9	5.1	6.4	5.2	4.6
IV Biothermic	5.0	3.7	7.4	6.5	5.3	6.2	5.8	5.6
V Composting plant	6.6	5.2	7.1	4.7	5.6	5.8	5.9	4.8
VI Landfill-4	6.5	9.2	8.8	2.4	4.7	6.9	6.4	6.2
VII Landfill-32	7.5	7.5	8.7	2.5	4.9	5.7	4.7	4.3
VIII Landfill & composting	7.1	4.9	7.5	3.7	6.7	5.8	5.3	3.5
				Lower Quartile				
I Incineration	2	8	8	6	2	5	4	7
II RDF combustion	4	6	7	5	4	5	5	6
III Fermentation	4	2	6	4	5	6	4	3
IV Biothermic	4	3	6	5	5	5	5	5
V Composting plant	6	4	6	4	5	6	5	4
VI Landfill-4	5	9	8	1	4	5	5	5
VII Landfill-32	6	5	8	1	3	5	5	4
VIII Landfill & composting	6	5	7	3	5	5	4	3

Alternative	g9 Resource Recovery	g10 Cost per ton
I Incineration	70	560
II RDF combustion	40	646
III Fermentation	35	708
IV Biothermic	40	790
V Composting plant	25	996
VI Landfill-4	15	455
VII Landfill-32	15	547
VIII Landfill & composting	20	620

There was relatively low variation in measures of alternative performance on criteria obtained from the questionnaire.

Multiple Attributes, Multiple Actors

The parties associated with the siting decision and the criteria of greatest importance to each are displayed below:

Parties and Criteria: Uusimaa Waste Disposal			
Environmental Criteria	Finnish Government	Technical Managers	Managers
Political feasibility	x		
Technical reliability	x	x	
Transport reliability	x	x	
Benefits to national economy	x		
Employment	x		
Acute environmental effects	x		x
Long-term environmental effects	x		x
Environmental hygiene	x		x
Resource recovery level	x	x	
Costs	x	x	

The Finnish Government, which motivated the development of local waste sites, considered all of the criteria. Local regions had technical managers responsible for proper disposal of waste, and environmental managers responsible to ensure environmental protection. The sequence of events in the decision process consisted of:

Sequence of Major Events: Uusimaa Waste Disposal

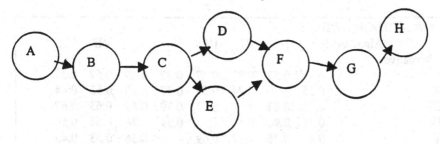

A   Finnish Waste Act of 1992
B   Uusimaa supervisory group formed
C   Criteria identified
D   Local technical managers surveyed for weights
E   Local environmental managers surveyed for weights
F   Panel of experts surveyed for measures of alternative performance
          over criteria
G   ELECTRE II analysis and sensitivity
H   Choice by supervisory group

Activity G in this process reflects the ELECTRE analysis directed by researchers.

ELECTRE II Ranking

Given the project impact matrix and weights of criteria importance, an outranking relationship based on a concordance test and a non-discordance test was constructed. The concordance relationships in Table 5 reflect the weighted proportion of criteria where the alternative indexed by column was viewed as preferable to the alternative indexed by row. (In Table 5, the column gives the weighted proportion of criteria where the column alternative is at least as good as the row alternative.) The discordance index for a pair of alternatives reflects the greatest proportional disadvantage between the compared alternatives. (In Table 5, the discordance index is for the column alternative compared to the row alternative.)

Table 5: Concordance & Discordance Matrices - Uusimaa Region

CONCORDANCE INDICES								
Alternative\Alternative	I	II	III	IV	V	VI	VII	VIII
I	-	0.37	0.37	0.37	0.37	0.69	0.57	0.37
II	0.76	-	0.24	0.57	0.37	0.69	0.53	0.44
III	0.76	0.83	-	0.77	0.79	0.63	0.63	0.67
IV	0.76	0.96	0.44	-	0.59	0.76	0.64	0.56
V	0.63	0.76	0.57	0.69	-	0.56	0.63	0.43
VI	0.54	0.61	0.37	0.61	0.57	-	0.63	0.44
VII	0.66	0.73	0.37	0.61	0.69	0.93	-	0.56
VIII	0.89	0.79	0.61	0.61	0.69	0.69	0.76	-
DISCORDANCE INDICES								
Alternative\Alternative	I	II	III	IV	V	VI	VII	VIII
I	-	0.55	0.70	0.59	0.82	1.00	1.00	0.91
II	0.31	-	0.47	0.35	0.34	0.66	0.66	0.42
III	0.47	0.16	-	0.13	0.28	0.50	0.50	0.27
IV	0.47	0.16	0.28	-	0.27	0.66	0.66	0.36
V	0.56	0.28	0.28	0.28	-	0.50	0.50	0.17
VI	0.42	0.35	0.82	0.70	0.59	-	0.49	0.47
VII	0.56	0.28	0.35	0.28	0.44	0.14	-	0.14
VIII	0.56	0.28	0.35	0.28	0.37	0.33	0.33	-

The outranking relationships were then used to analyze the multicriteria selection problem. A strong and a weak outranking relationship were both defined. A **strong outranking** between a pair of alternatives is defined as satisfying one of two conditions:

a concordance index greater than or equal to a limit p^*, a discordance index no greater than q^*, and the sum of weights where the base alternative has an advantage is at least as great as the sum of the weights where the other alternative has an advantage.

Or, a concordance index greater than or equal to a limit p^0, a discordance index no greater than q^0, and the sum of weights where the base alternative has an advantage is at least as great as the sum of the weights where the other alternative has an advantage.

Parameter p^* is a tighter specification than parameter p^0, and parameter q^* is a looser specification than parameter q^0. The results of the outranking relationships will be sensitive to different parameter settings, so 665 different combinations of parameter values were used within the following ranges in increments of 0.05.

$$0.70 \leq p^* \leq 0.90$$
$$0.65 \leq p^0 \leq 0.85$$
$$0.60 \leq p^- \leq 0.80$$
$$0.40 \leq q^* \leq 0.55$$
$$0.20 \leq q^0 \leq 0.40$$

In all cases, $p^* > p^0 > p^-$ and $q^* > q^0$, and $p^* - p^0 \geq 0.05$, $p^0 - p^- \geq 0.05$, and $q^* - q^0 \geq 0.05$.

The **weak outranking** relationship is defined as a concordance index greater than or equal to a limit p^-, a discordance index no greater than q^*, and the sum of weights where the base alternative has an advantage is at least as great as the sum of the weights where the other alternative has an advantage. The strong outranking relationships are given in Table 6, and the weak outranking relationships in Table 7 using values of $p^* = 0.7$, $p^0 = 0.65$, $p^- = 0.6$, $q^* = 0.5$, and $q^0 = 0.3$.

Table 6: Strong Outranking Relationships - Uusimaa Region

STRONG RELATIONSHIP				
Alternative I	outranks	Alternative II Alternative III Alternative IV	outranked by	nothing
Alternative II	outranks	Alternative III Alternative IV Alternative V Alternative VII Alternative VIII	outranked by	Alternative I
Alternative III	outranks	nothing	outranked by	Alternative I Alternative II Alternative IV Alternative V
Alternative IV	outranks	Alternative III	outranked by	Alternative I Alternative V Alternative II
Alternative V	outranks	Alternative III	outranked by	Alternative II Alternative IV
Alternative VI	outranks	Alternative VII	outranked by	nothing
Alternative VII	outranks	Alternative VIII	outranked by	Alternative II Alternative VI
Alternative VIII	outranks	nothing	outranked by	Alternative II Alternative VII

Table 7: Weak Outranking Relationships - Uusimaa Region

WEAK RELATIONSHIP			
Alternative I outranks	Alternative II Alternative III Alternative IV	outranked by	nothing
Alternative II outranks	Alternative III Alternative IV Alternative V Alternative VI Alternative VII Alternative VIII	outranked by	Alternative I
Alternative III outranks	nothing	outranked by	Alternative I Alternative II Alternative IV Alternative V Alternative VI Alternative VII
Alternative IV outranks	Alternative III Alternative VII Alternative VIII	outranked by	Alternative I Alternative II
Alternative V outranks	Alternative III Alternative VIII	outranked by	Alternative II
Alternative VI outranks	Alternative III Alternative VII Alternative VIII	outranked by	Alternative II
Alternative VII outranks	Alternative III Alternative VIII	outranked by	Alternative II Alternative IV Alternative VI
Alternative VIII outranks	nothing	outranked by	Alternative II Alternative IV Alternative V Alternative VI Alternative VII

Graphs were constructed for both the strong and weak relationships. An upward ranking was obtained by identifying each alternative not outranked on the graph, and placing those alternatives in the first rank. The outrankings involving assigned alternatives were deleted, and this process continued until all alternatives were ranked. A downward ranking was obtained by applying the same process in reverse. This was done for both strong and weak relationships. Table 8 gives the results of these rankings, which led to the final ranking given in Table 9.

Table 8: ELECTRE II Rankings – Uusimaa Region

	STRONG RELATIONSHIP		WEAK RELATIONSHIP	
	Upward	Downward	Upward	Downward
First rank	I, VI	I	I	I
Second rank	II	II	II	II
Third rank	IV, VII	IV, VI	IV, V, VI	IV, VI
Fourth rank	V, VIII	V, VII	VII	V, VII
Fifth rank	III	III, VIII	III, VIII	III, VIII

Ranks by alternative					Final
I	1	1	1	1	1
II	2	2	2	2	2
III	5	5	5	5	5
IV	3	3	3	3	3
V	4	4	3	4	4
VI	1	3	3	3	3
VII	3	4	4	4	4
VIII	4	5	5	5	5

Table 9: ELECTRE II Ranking - Uusimaa Region

		Sites	Method
Group 1	Alternative I	1	Incineration
Group 2	Alternative II	4	RDF combustion
Group 3	Alternative IV	4	Biothermic
	Alternative VI	4	Landfill
Group 4	Alternative V	4	Composting plant
	Alternative VII	32	Landfill
Group 5	Alternative III	4	Fermentation
	Alternative VIII	32	Landfill & composting

Sensitivity

Because ELECTRE requires a large number of parameters to be set, different values of the p and q parameters were used, in intervals of 0.05. This resulted in 665 combinations of parameters. Among all of these combinations, there were a minimum of three clusters, and a maximum of eight (which is the maximum number of clusters possible, as there were eight alternatives). Centralized incineration (alternative I) was ranked in the first cluster 646 times out of the 665 combinations. When uncertainty in criteria weights was considered, no differences appeared in the rankings of the best alternatives.

Oulu Region Study

This study was similar to the Uusimaa region study, but applied a later development of the outranking approach, ELECTRE III (Hokkanen and Salminen, 1997a). An outranking method was selected because there were 113 people involved in the decision, each with their own preference structure. The task was to produce an acceptable compromise solution. The inability to elicit preference information in detail precluded the use of multiattribute utility theory or the analytic hierarchy process. The presence of imprecise data made ELECTRE III preferable to ELECTRE II. While the general idea of ELECTRE III is the same as that of ELECTRE II, ELECTRE III is expected to be less susceptible to variations in data and parameters. The decision problem involved a planning region of 17 municipalities in northern Finland serving a population of about 185,000 people. The decision was to select a municipal solid waste management system to process about 80,000 tons per year until the year 2010. About 15,000 tons of this material was recyclable, primarily paper and cardboard. At the time of the study, each municipality dealt with its own waste, but their procedures did not satisfy the Finnish Waste Act of 1992. Existing landfills were unstaffed and unmonitored, with the potential to release pollutants.

Treatment methods considered were sanitary landfill, incineration, and composting. Incineration involved used of refuse-derived fuel when possible, and peat when necessary. In addition to the treatment method, options reflected decentralized (seventeen sites), centralized (one facility for the region), and intermediate systems. Alternatives are given in Table 10.

Table 10: Solid Waste Disposal Alternatives for the Oulu Region

Treatment Method	Cooperation Level	Number of Sites Landfill	Compost	RDF
IA Landfill	Decentralized	17		
IB Landfill and open composting	Decentralized	17	17	
IC Landfill, composting, RDF	Decentralized	17	17	1
IIAa Landfill	Intermediate	4		
IIAb Landfill and open composting	Intermediate	4	4	
IIAc Landfill, composting, RDF	Intermediate	4	4	1
IIBa Landfill	Intermediate	4		
IIBb Landfill and open composting	Intermediate	4	4	
IIBc Landfill, composting, RDF	Intermediate	4	4	1
IICa Landfill	Intermediate	3		
IICb Landfill and open composting	Intermediate	3	3	
IICc Landfill, composting, RDF	Intermediate	3	3	1
IIDa Landfill	Intermediate	6		
IIDb Landfill and open composting	Intermediate	6	6	
IIDc Landfill, composting, RDF	Intermediate	6	6	1
IIEa Landfill	Intermediate	4		
IIEb Landfill and open composting	Intermediate	4	4	
IIEc Landfill, composting, RDF	Intermediate	4	4	1
IIIA Landfill	Centralized	1		
IIIB Landfill centralized, compost	Decentralized	1	17	
IIIC Landfill, compost, RDF	Centralized	1	1	1
IIID Landfill & RDF	centralized exc. com	1	17	1
IIA, IIB, and IIE involved different combinations of cooperating municipalities				

The decision-makers were the municipal councils involved, after hearing reports from environment and technical committees subordinate to them. Potential interest groups participated in a supervisory group overseeing the project.

Objectives

A preliminary list of potential objectives was generated from four sources: objectives derived from functional elements, from the literature, those stated in the Waste Act, and those added by decision-makers. A list of 28 measures were generated. A group of 113 members of the environment and technical committees of the 17 municipalities participated in aggregating these objectives into a comprehensive, operational, nonredundant and minimal set of criteria. The eight criteria selected are given in Table 11.

Table 11: Weights Based on Questionnaire Responses – Oulu Region

| | | | For sensitivity analysis | |
		Majority	Min used	Max used
g1 Net cost per ton	economic	0.27	0.25	0.35
g2 Technical reliability	technical	0.26	0.20	0.30
g3 Global environmental effects	environmental	0.016	0.02	0.39
g4 Local and regional health effects	environmental	0.096	0.01	0.20
g5 Acidificative releases	environmental	0.047	0.005	0.04
g6 Surface water dispersed releases	environmental	0.12	0.11	0.18
g7 Number of employees	employment	0.05	0.04	0.08
g8 Amount of recovered waste	recovery	0.14	0.06	0.17

Some of these criteria may still be highly correlated. Because of the complexity of the relationships among variables, all strong correlations between criteria could not be avoided.

Measures

Cost (g1) was measured as the estimate of total annual cost in Finmarks per ton of waste processed, minus revenues. Costs included collection and transport. Technical reliability (g2) was assessed by expert questionnaire with values ranging from 0 to 10. Four environmental criteria (g3 through g6) were included. Global effects (g3) represented the total greenhouse effects of each alternative, and were measured in a weighted total of gas released per kilogram of waste. Health effects (g4) measured lead, cadmium, arsenic, and mercury releases to air and water in milligrams per kilogram of waste processed. Acidificative releases (g5) estimated sulphur dioxides and nitrogen oxides emitted. Surface water dispersed releases (g6) was represented by nitrogen concentrations in leachates. The number of employees required was considered a positive feature, so criterion g7 was to be maximized. Resource recovery level (g8) was the sum of paper, organic, and fuel wastes recovered. Measures for the 22 alternatives evaluated are given in Table 12.

Table 12: Project Impact Matrix - Oulu Region

Alternative	g1 Cost Min	g2 tech reliab Max	g3 global effects Min	g4 health effects Min	g5 acid release Min	g6 water release Min	g7 empl Max	g8 resource recovery Max
IA	656	5	552,678,100	609	1190	670	14	13900
IB	786	4	539,113,200	575	1190	682	18	23600
IC	912	4	480,565,400	670	1222	594	24	39767
IIAa	589	9	559,780,715	411	1191	443	10	13900
IIAb	706	7	532,286,214	325	1191	404	14	23600
IIAc	834	6.5	470,613,514	500	1226	384	18	40667
IIBa	580	9	560,987,877	398	1191	430	10	13900
IIBb	682	7	532,224,858	314	1191	393	14	23600
IIBc	838	6.5	466,586,058	501	1229	373	22	41747
IICa	579	9	561,555,877	373	1191	405	9	13900
IICb	688	7	532,302,258	292	1191	370	13	23600
IICc	838	6.5	465,356,158	499	1230	361	17	42467
IIDa	595	9	560,500,215	500	1191	538	12	13900
IIDb	709	7	532,974,014	402	1191	489	17	23600
IIDc	849	6.5	474,137,314	648	1226	538	20	40667
IIEa	604	9	560,500,215	500	1191	538	12	13900
IIEb	736	7	532,974,014	402	1191	489	17	23600
IIEc	871	6.5	474,137,314	648	1226	538	20	40667
IIIA	579	9	568,674,539	495	1193	558	7	13900
IIIB	695	6	536,936,873	424	1195	535	18	23600
IIIC	827	7	457,184,239	651	1237	513	16	45167
IIID	982	7	457,206,173	651	1239	513	16	45167

Multiple Attributes, Multiple Actors

The actors at Oulu were similar to those at the Uusimaa site. However, each locale considered different sets of criteria, important to them.

Parties and Criteria in the Oulu Waste Disposal Site Decision			
Criteria	Finnish Government	Technical Managers	Environmental Managers
Cost	x	x	
Technical reliability	x	x	
Global environmental effects	x		x
Local and regional health effects	x		x
Acidicative releases	x		x
Surface water dispersed releases	x		x
Number of employees	x	x	
Amount of recovered waste	x	x	

The sequence of events was similar to that of the Uusimaa decision.

Sequence of Major Events: Oulu Waste Disposal

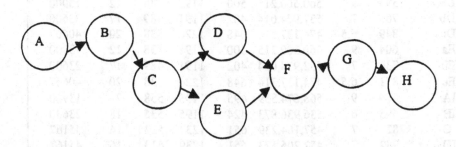

A   Finnish Waste Act of 1992
B   Uusimaa supervisory group formed
C   Criteria identified
D   Local technical managers surveyed for weights
E   Local environmental managers surveyed for weights
F   Panel of experts surveyed for measures of alternative performance
         over criteria
G   ELECTRE III analysis and sensitivity
H   Choice by supervisory group

The ELECTRE method applied, however, was different from the Uusimaa case.

Weights and Parameters

Weights were generated from technical and environmental committees of the municipalities. Two methods were used to determine weights. First, the 113 participants were asked to assign weights ranging from 1 to 7 to each criterion. The second method was to have each of the participants assign 1 to the least important criterion, and assign values proportional in importance to the others. Average and median weights were only slightly different across these two procedures. The first approach was the basis for the weights used, with the final weight being the highest (or lowest if minimizing) that would obtain a majority. (The 113 weights for each criterion were rank ordered, and the 57th ranked weight used, as shown in Table 11.) ELECTRE III requires preference, indifference, and veto thresholds (Roy and Vincke, 1984; Vincke, 1992).

ELECTRE Ranking

The initial analysis resulted in the partial order given in Table 13. There were 13 groups, with alternative IIBc (decentralized landfill combined with open composting) in the top-ranked group by itself.

Table 13: ELECTRE III Ranking - Oulu Region

Alternatives		Landfill	Composting	RDF Incineration
Group 1	IIBc	4 sites	4 sites	1 site
Group 2	IIAc	4 sites	4 sites	1 site
	IICb	3 sites	3 sites	
Group 3	IIBb	4 sites	4 sites	
Group 4	IICc	3 sites	3 sites	1 site
	IIDa	6 sites		
	IIEa	4 sites		
Group 5	IIIB	1 site	17 sites	
Group 6	IIIC	1 site	1 site	1 site
Group 7	IIAa	4 sites		
	IIBa	4 sites		
	IIAb	4 sites	4 sites	
	IIDc	6 sites	6 sites	1 site
Group 8	IICa	3 sites		
	IIDb	6 sites	6 sites	
	IIEc	4 sites	4 sites	1 site
Group 9	IIEb	4 sites	4 sites	
Group 10	IIIA	1 site		
Group 11	IIID	1 site	17 sites	1 site
Group 12	IA	17 sites		
Group 13	IB	17 sites	17 sites	
	IC	17 sites	17 sites	1 site

Sensitivity Analysis

Because of the uncertainty involved in the measures, a sensitivity analysis was conducted on both criteria weights and on the threshold values used. When weights were varied, the highest ranked alternatives remained in the same order. Changing preference and indifference threshold values did not change the sole first-rank of alternative IIBc. When the preference threshold for surface water dispersed was given a very low value, alternative IIAc was the sole alternative in the second-ranked group. There was considerable change when veto threshold values, used in generating the discordance index, were varied. When discordance was in effect removed from the model, alternative IIBb was ranked first, while alternatives with regional landfills were ranked in a second group. This was due to the resource recovery level. Since the Finnish Waste Act specifically included the goal of higher resource recovery, the original solution (IIBc) was adopted.

Post-Study Success

The participants were surveyed for their opinions about the process. The responses indicated positive attitudes, emphasizing that they did not have much time to devote to the task. The method was also viewed positively in that it seemed to the participants to minimize political influence of individual participants on the process.

The situation prior to the study was that only four or five of the existing 24 landfills had been operating in compliance with standards. Comparison of the selected alternative with the existing situation found savings of 60 million Finmarks up to the project end-time of 2010. Recovered wastes would be increased, and there would be fewer pollutants.

Kirkkonummi Study

Kirkkonummi is a Finnish municipality in the southern area of the country. The Finnish Parliament had passed a Building Act in 1994 that suggested municipalities develop a general land-use plan for environmental processing of waste through year 2020 (Hokkanen, et al., 1998). Decision-makers were members of the Planning Board of the municipality, along with other city officials.

Generation of Alternatives

The planning process involved Planning Board meetings to define the problem, identify regional alternatives, identify environmental impacts, define criteria, obtain measures of alternative performance on criteria, conduct a multicriteria analysis, and develop a suggested order of preference among alternatives for the Municipal Council. The process involved some redefinition in that alternatives were redefined after the initial multicriteria analysis.

The original six alternatives were to place the disposal site at one of six locations (the Centre, Kolsarby, Kantvik, Masala, Sundsberg, and Veikkola). The learning process involved understanding that the original Sundsberg seashore area would not be an acceptable site. Alternative VII was a joint adoption of the Kolsarby (II), Kantvik (III), and Masala and Sundsberg North (IV) alternatives. The final set of alternatives considered were the seven alternatives given in Table 14.

Table 14: Kirkkonummi Planning Data

Criteria Alternative	g1 effects on Environment Max	g2 economics Min	g3 land ownership Max	g4 community structure Max	g5 health Min
I The Centre	5	1490	15	4.0	42
II Kolsarby	2.5	3780	25	2.5	14
III Kantvik	3	2750	10	2.8	0
IV Masala & Sundsberg N	3	3220	16	3.2	13
V Sundsberg S & Sarvvik	2	6720	0	1.0	22
VI Veikkola	4	3440	30	3.5	6
VII joint alternative	.3	3310	17	3.0	9

Criteria Selection

All effects required by the motivating legislation were considered, along with urban planning economics. In addition, the participants generated dozens of additional factors. However, of these additional factors, only land ownership was not overlapping with the original environmental and economic factors. Eight preliminary criteria resulted. Irrelevant criteria, such as effects on landscape, effects on cultural history, and utilization of natural resources, were removed. Five final criteria resulted, as given in the column headings of Table 14.

Initially, participants that would have been adversely affected from the selection of a particular alternative were highly critical of the criteria and their measures. The multicriteria approach was helpful because the measures were clearly defined. This resulted in agreement on criteria and measures, and focus on the differences between alternatives.

Multiple Attributes, Multiple Actors

The criteria most important to the parties involved were:

Parties and Criteria for the Kirkkonummi Study			
Environmental Criteria	Finnish Government	Technical Managers	Managers
Effects on the environment	x		x
Economics	x	x	
Land ownership	x	x	
Community structure		x	x
Health	x	x	x

Alternative Measures on Criteria

The impact on the environment was measured on a five point ratio scale, with 5 representing the best condition of minimal effects on natural ecosystems and 1 representing significant changes to ecosystems. The Centre had the highest value because construction would be in an urban environment, resulting in no changes in the stress to the ecosystem. Southern Sundsberg and Sarvvik was still in its natural state, and thus received a low rating.

Economic impact was measured in Finmark impact per inhabitant of the region. Costs for construction, maintenance, water supply, and road networks were included. Cost differences across alternatives depended on ground properties and connections to service networks. Those costs covered by the Finnish government were not included.

Land ownership was measured by percentage of the area owned by the municipality. A larger proportion was preferred because it was expected to make the plan easier to implement.

Impact on community structure was assessed on a five point ratio scale. The municipality had decided that new construction should support the current community structure. The ratings on this criterion were closely correlated with the ratings given on the criterion of impact on environment.

Impact on health and living conditions was measured based on the area where the noise level from the road network exceeded 55 decibels.

Multicriteria Analysis

ELECTRE IV was not used in this analysis because of tendencies of that approach to favor cautious alternatives that do not have any extreme measures (good or bad), and a reported tendency to overweight criteria that were easier to measure (Hokkanen and Salminen, 1997b). ELECTRE III concordance indices were used, combined with the stochastic multiobjective acceptability analysis (SMAA) method (Lahdelma, Hokkanen and Salminen, 1998). The SMAA method generalizes the efficiency concept when inaccurate data and partial preferences are represented as probability distributions. The weight space ideas of SMAA were combined with the pseudo criteria threshold model of ELECTRE III. Discordance indices were not used, nor were weights required for the criteria.

The general sequence of the decision process was similar to that in the prior two studies.

Sequence of Major Events: Kirkkonummi Waste Disposal

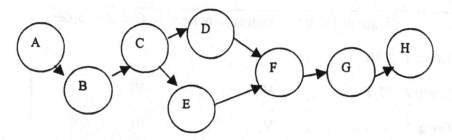

A   Finnish Building Act of 1994
B   Uusimaa supervisory group formed
C   Criteria identified
D   Local technical managers surveyed for weights
E   Local environmental managers surveyed for weights
F   Panel of experts surveyed for measures of alternative performance over
        criteria
G   ELECTRE rankings and sensitivity
H   Tentative decisions

Results of Analysis
Acceptability indices, central weight vectors, and weight bounds were
computed for each alternative, following SMAA. In the first iteration,
alternatives II and V were ranked lower than the other five. Considering
that there was a binary option between (II, III, and IV) or (VII) because
these were two different ways to do the same work, the analysis resulted in
alternative implementation orders given in Table 15.

Table 15: Kirkkonummi Rankings of Alternatives

	Suggested order I	Suggested order II	Complete order
Group 1	I, VI	I, VI	I
Group 2	III, IV	VII	VI
Group 3	II	V	III
Group 4	V		IV
			II
			V

In this study interest groups had clear opinions of alternative sites. Those conflicts due to misunderstanding were reduced by showing impact of construction of each alternatives. Conflicts based on differing interests were dealt with by identifying those who benefited explicitly. Conflicts based on different valuation of factors was dealt with by providing the possible implementation orders obtained from SMAA. This output gave descriptive information in terms of which weight emphases would support each alternative.

Originally, participants were skeptical. Presentation of results generated spirited conversations. A tentative decision was made to adopt sites at Kantvik, Masala and Sundsberg North, and Kolsarby (alternatives II, III, and IV). Annual revisions to the decision were allowed to consider possible additional information.

Conclusions

These three studies show different implementations of the outranking method ELECTRE. There was a fourth study, applied at Jämsä, Finland (Hokkanen and Salminen, 1997b) reported by those authors to compare ELECTRE III and ELECTRE IV. ELECTRE IV does not consider weights of relative importance, but simply counts the number of criteria where each alternative has a relative advantage. There are different methods used to determine outranking as well. Experimenting with the results of the Oulu and Jämsä studies, ELECTRE IV was found by Hokkanen and Salminen to yield quite different rankings at the less important groupings of alternatives. For the top ranked alternative, ELECTRE IV yielded the same result as ELECTRE III in the Oulu study, but ranked alternative 5 in the 7th group and alternative 8 in the second group, as opposed to those alternatives being ranked in the first group in the ELECTRE III study. ELECTRE III was judged by Hokkanen and Salminen to be more appropriate for environmental studies than ELECTRE IV. Salminen, et al. (1998) compared ELECTRE III, PROMETHEE I, PROMETHEE II, and SMART for these four applications, as well as on a randomly generated set of problems. The methods were found to generate quite different sets of recommended alternatives in some cases. Salminen, et al. recommended using several methods whenever possible. They recommended that if only one method could be applied, that ELECTRE III be selected over the others.

Rogers and Bruen (1999) applied ELECTRE I, ELECTRE III, and ELECTRE IV to three city solid waste management decisions in Ireland. ELECTRE I was applied to a Galway City decision with qualitative data. ELECTRE III was used for selecting a motor vehicle road through Dublin, with more detailed data (both quantitative and qualitative). ELECTRE IV was used to rank nine alternative landfill sites for East Wicklow. The East Wicklow case involved detailed qualitative data with no basis for deducing relative weights.

Rogers and Bruen (1998a) argued that the ELECTRE III threshold parameters p (the preference threshold reflecting the amount of relative advantage at which the decision-maker shows a clear strict preference on a criterion), q (the indifference threshold beneath which the decision-maker is indifferent to relative performance on a criterion), and v (the veto threshold where this degree of difference on a criterion would negate any possible outranking on other criteria) all required careful selection based on the context of the problem. These same authors (1998b) published a study using ELECTRE III. They proposed a system of weight elicitation similar to swing weighting (von Winterfeldt and Edwards, 1986).

The outranking methods provide a family of models that can be applied for different types of data. Both the Finnish and Irish applications found that different models were best for different problems based on varying data availability.

References

Hokkanen, J., Salminen, P., Rossi, E. and Ettala, M. The choice of a solid waste management system using the ELECTRE II decision-aid method. *Waste Management & Research* 13, 1995, 175-193.

Hokkanen, J. and Salminen, P. Choosing a solid waste management system using multicriteria decision analysis. *European Journal of Operational Research* 98, 1997a, 19-36.

Hokkanen, J. and Salminen, P. ELECTRE III and IV decision aids in an environmental problem. *Journal of Multi-Criteria Decision Analysis* 6, 1997b, 215-226.

Hokkanen, J., Lahdelma, R., Miettinen, K. and Salminen, P. Determining the implementation order of a general plan by using a multicriteria method. *Journal of Multi-Criteria Decision Analysis* 7, 1998, 273-284.

Lehdelma, R., Hokkanen, J. and Salminen, P. SMAA – Stochastic multiobjective acceptability analysis. *European Journal of Operational Research* 106, 1998, 137-143.

Rogers, M. and Bruen, M. Choosing realistic values of indifference, preference and veto thresholds for use with environmental criteria within ELECTRE. *European Journal of Operational Research* 107, 1998a, 542-551

Rogers, M. and Bruen, M. A new system for weighting environmental criteria for use within ELECTRE III. *European Journal of Operational Research* 107, 1998b, 552-563.

Rogers, M. and Bruen, M. Applying ELECTRE to an option choice problem within an environmental appraisal – Three case studies from the Republic of Ireland. In *Advances in Decision Analysis*, N. Meskens and M. Roubens, eds., Kluwer Academic Publishers, Dordrecht, 1999, 159-200.

Roy, B. and Bouyssou, D. *Aide multicritere a la decision: Methodes et cas.* Collection Gestion, Economica, Paris, 1993.

Roy, B. and Vincke, P. Relational systems of preference with one or more pseudo-criteria: Some new concepts and results. *Management Science* 11, 1984, 1323-1335.

Salminen, P., Hokkanen, J. and Lahdelma, R. Comparing multicriteria methods in the context of environmental problems. *European Journal of Operational Research* **104**:3, 1998, 485-496.

Vincke, P. *Multicriteria Decision-Aid*, Chichester: Wiley, 1992.

von Winterfeldt, D. and Edwards, W. *Decision Analysis and Behavioral Research*, New York: Cambridge University Press, 1986.

7 PIPELINE LOCATION DECISIONS

Pipelines provide a very important means of transporting the fuels that make civilization prosperous. Pipelines are relatively safe and an efficient means to ship gas and liquid petroleum products. Dey, et al. (1998) stated that pipelines have failure rates much lower than railroads or highway transportation, although failures do occur. In 1993, 51 people were burnt to death when escaping gas from a pipeline ignited. In 1994, a pipeline in New Jersey (USA) failed, killing one and injuring over fifty other people. Pipeline failures have occurred around the world. These failures rarely result in fatalities, but lead to heavy business losses, and cause considerable damage to the environment.

Pipeline location decisions have a number of common characteristics:

1. They involve safety and security issues, because there is a long history of pipeline accidents. These accidents sometimes involve undesirable waste in the form of spills, but also have involved explosions with loss of life.

2. Pipeline routes are complicated in that they are rarely located as the shortest route connecting two points, but usually involve routing around blockages, as well as over locales with better soil support. The identification of the initial route is a series of local site selection

decisions.   This is especially true when pipelines are located in populated areas. Laws in Russia, as well as the U.S., regulate limits on possible pipeline routes in order to obtain a high level of human safety and conservation of the environment.

3. Pipelines involve choices at the design stage, as there are a number of different pipeline options to be determined in the original design.

4. Each pipeline siting is unique, although all pipelines involve some common features. Therefore, examination of problems on one pipeline provide valuable information for the next pipeline, but each project will have a unique set of problems as well. This creates the need for new selection decisions.

5. Pipeline variants involve high levels of uncertainty. Each alternative involves unique characteristics, with incomplete knowledge about route performance.

Pipeline location problems have many of the features common to other location decisions. First, there are usually several active groups with strong interests in the decision. There are a number of criteria over which alternative sites can be evaluated. The different groups have different perceptions of the choice problem. Usually, every interest group have their own subset of criteria which they view as important. The Kunreuther and Linnerooth multiple attribute multiple party model is especially appropriate for this class of siting decision.

We will present three different pipeline route location decisions. These cases were in different countries, over different periods of time. For each case, we will describe:

1. the case, including the problem, the organization responsible for solving the problem, and the active interest groups.

2. We will review the criteria important to the decision-maker, as well as to the interests groups involved. Where known, we will describe how each alternative performed on each of these criteria.

3. We will describe the analytic models used in each case.

The chapter will conclude with evaluation of common features of the three cases. We will attempt to define the objective requirements for an analytic model in order to deal with this class of decision problem.

CASE 1: Pipeline Route Selection in the Southern Soviet Union

Large natural gas deposits are important for the development of Russia. This was true in the Soviet Union as well. Gas is the cheapest source of energy for use in industry, agriculture, and support of everyday living.

Decision Problem

In the former Soviet Union, there were plans for construction of new pipelines in different regions of the country. Plans included a new pipeline route from the southern Soviet Union (now the independent state of Georgia). The decision problem was selection of a route crossing a heavily populated area over valuable agricultural land.

Construction of a gas pipeline involved a zone of 250 to 350 meters from the centerline of the route needed to buffer residential buildings, highways, farms, and other installations. The pipeline transported gas about one thousand kilometers. The buffer area was thus 50 thousand to 70 thousand hectares in size. Therefore, the pipeline route had a major impact on the region that it crossed, making it a very important decision.

There also was significant impact on populated areas. This made it necessary to have thicker pipe, in order to provide added assurance against bursting. This used up valuable metal. It also increased construction costs, along with demolition of buildings and dwellings where necessary. The pipeline also had a long-term impact on development of populated areas and cities for about 25 years.

Funds were required to cover reimbursement of losses due to pulling valuable agricultural land out of production. There also was added work required to recultivate replacement land and restore irrigation systems.

The specific problem was to build a gas pipeline route connecting the cities of Kutaisi and Sukhumi, with branch pipelines to Poti and Batumi to supply gas to domestic and industrial consumers. Further details are given in Oseredko, Larichev, and Mechitov (1982).

Alternatives

The initial stage of the analysis involved research, field inspections, and preliminary agreements. Three possible routes were identified. The first route crossed a piedmont area, the second partially crossed piedmont and partially crossed water, and the third route was primarily maritime.

The **piedmont route** was the shortest pipeline, crossing spurs of the Egriss ridge. The relief of this path is heavily criss-crossed by canyons carrying mountain rivers. Elevation varies about 700 meters over the piedmont route. There were a large number of small villages along this route, requiring demolition of many homes. Avoiding this demolition would require a complicated route in mountainous terrain. The piedmont route involved construction through muddy torrents with karst land forms and landslides, aggravated by the need to cut special terraces into steep slopes in order to move construction machinery and to lay the pipeline. The mountain route was at some distance from heavily developed areas, so that there was not good road support for construction. Pipeline surveillance and maintenance would only be possible through the use of helicopters.

The **median version** passed through more populated zones. The relief for this route was gentle, with good geological conditions and well-developed road and rail systems. However, this path would require the greatest amount of building demolition and destruction of cultivated areas. It would also require numerous crossings over artificial obstacles.

The **maritime route** was the shortest path, passing through the Kolkhida lowland area. The relief is fairly flat, with considerable areas of alder woodland and a well-developed irrigation system. This route risked some crossing through swamps, making conditions uncertain, especially during rainy periods. The maritime route would require very little building demolition or crop destruction. However, it would require some branch pipelines that would result in demolition and crop loss. In the preliminary study, the area was described through topographic maps drawn from aerial photographs. Therefore, detailed information was lacking in the initial stages of the analysis.

Conventional Practice

The most common and widely used estimation criterion for projects in the former Soviet Union was the model:

$$Cost = Kx + A$$

where K is the capital investment, x is the normative coefficient of capital investment efficiency (for industrial construction, this was 0.12), and A is the annual maintenance cost. This basic criterion of value permitted selection of a route from an initial to a final point requiring the minimum total capital outlay in equipment and labor, as well as maintenance expense. However, this approach did not guarantee the selection of the truly optimal route, because not all environmental and social factors were included, and because of the high uncertainty in estimating future costs, and transforming these costs into net present terms (Goncharov and Oseredko, 1977).

Construction time is one of the most decisive factors in determining the viability of a gas pipeline. Construction duration is partially dependent on the standards required, but also can be affected by mandated directions from the government. In general, the preferred route alternative is that route where construction organizations already exist and where seasonal transport routes are available, or where pipelines already exist together with appropriate maintenance systems. The factor of minimum change in existing construction technology and machinery is also an important factor, as well as the availability of an adequate labor force.

Gas Pipeline Maintenance: In order to ensure reliable pipeline operation, it is necessary to have access to all sections for preventive inspections and repair work should pipeline components fail. Access is to some extent determined by environmental conditions in the pipeline area and by the development of the transportation system.

Maintenance reliability primarily depends on natural climatic conditions along the route. In some cases, to ensure faultless operation of gas transportation systems in the most complex and important sections (some involving large areas of water, swamps, or inaccessible mountain regions), two pipelines are laid rather than one as a matter of standard practice. This increases capital investment, of course, and can also create the need for gas reservations such as underground storage in natural geological formations.

Environmental Impact: Construction of primary gas pipelines, especially those of one thousand kilometers or longer, has obvious environmental impact. This scale of construction involves partial forest clearance and inevitable loss of agricultural production. The long-term environmental impacts are often overlooked in project evaluation.

When laying a pipeline in highlands, pipeline reliability is affected since construction can induce landslides. Hydrological systems can also be disrupted.   For instance, construction of underground pipelines at insufficient depth can deform water channels and disrupt drainage that can adversely affect aquatic life and disrupt pipeline operation.

Pipeline installation can also generate other environmental damage. Noise from gas turbine units can seriously influence maintenance personnel and local inhabitants, as well as animal life.

Relationship to Regional Development Plans: Gas pipeline construction also impacts the population and economics of the region through which it passes. Construction of large pipelines requires several thousand specialists and workers, some from other countries. The type of labor force required during different phases of construction changes sharply. The socio-economic influences of this influx of labor on local populations can be important, and needs to be considered in evaluation. If demolition of homes is necessary, new problems of rehousing and development of new residential areas are created. It also is important to provide social and cultural facilities and other services for maintenance personnel and their families. This can generate significant additional capital cost.

Cost factors involved in the construction of pipeline alternatives can sometimes be compensated in part by the benefits of supplying the area with gas, a new resource that can provide centralized heating systems for homes. Other benefits include use of waste hot water from compressor station gas turbine cooling systems, the new socio-cultural and life-enhancing facilities developed for the pipeline crews, communications systems, and other facilities.

New gas pipelines can have a major influence on local public opinion. The attitude of the local administrative bodies and the local population toward approaching construction need to be considered. Positive measures to obtain local support can be invaluable in making it possible to reallocate land use, obtain permission for various construction activities, and speed completion of design and surveying work, as well as of construction itself. Local support obviously improves the economic viability of the project.

Construction Conditions: The amount of construction effort required is heavily influenced by geological, hydrological, topographic, and other conditions along the selected pipeline route. Availability of existing infrastructure and construction bases is also important. Construction conditions need to be isolated as an independent factor because they are important for the construction firm that also participates in the route selection process.

Population Safety is ensured primarily through maintaining standard minimum distances from the main gas pipeline axis to populated areas, buildings, farms, highways, and other facilities. However, this safety buffer does not completely guarantee safety of the population in emergency situations. Potential risks can be reduced in two ways:

(i)   by increasing the reliability of technological systems and installations;

(ii)   by extending the gas pipeline guarding zone and distances from the various systems to the nearest populated areas.

Regulations concerning violations of the guarding zone, and fire protection regulations in the former Soviet Union were the most extensive in the world. However, this made pipeline projects more expensive.

Special Permissions

Special permission requirements and limitations imposed by laws to protect nature, fish reserves, mineral resources, forestry, and other entities were needed. Compliance with sanitary and fire protection authority regulations were also required. These regulations restricted possible pipeline routes.

Evaluation of Alternatives Over Criteria

A commonly used approach to evaluate alternatives over multiple criteria is:

a) to develop an evaluation scale for each criterion. Usually these are continuous numeric scales, or discrete numeric scales.

b) To calculate evaluations for each alternative on the basis of output from a mathematical model, from formulas, or from subjective evaluation by an expert.

Unfortunately, this approach was not applicable in the pipeline problem being described. The decision making body considered only one criterion – net present cost. There is an objective quantitative scale for evaluation using the formula presented earlier that could be applied to all three alternatives.

Other criteria that could be important were not readily measurable in quantitative form because of their subjective nature. The first attempt was to develop a qualitative scale using verbal descriptions of different grades of quality for other criteria. Although this could be done (for example, assigning evaluations like "good" and "bad" for population safety), experts refused to provide the needed estimates. This was because they lacked experience for projects of the scale being analyzed, and each large project has unique features that limit transference of such experience had it been available. Therefore, experts were asked for comparative judgments about the quality of each alternative on each criterion. The experts were able to provide this level of judgment based on past experience.

Alternative Costs: Construction costs for the three alternatives evaluated depended upon the difficulties encountered in laying gas pipeline over different terrain (swamps, river crossing, on terraced mountain slopes, etc.) and maintenance costs. Additional costs arise from building demolition, and for compensation for items such as crossing cultivated land. Local administrators estimated these costs. According to the formula used, the three alternatives had costs of:

> Piedmont route 10.8 million rubles
> Median route   8.5 million rubles
> Maritime route   8.9 million rubles

The pipeline construction process can be divided into two stages. The first stage is construction of the main pipeline itself. The second stage is laying branch pipelines (from the main pipeline to consumers). Not only was this work done at different times, but it was also funded from different sources. Branch pipelines were paid for by regional budgets. Therefore, two cost criteria had to be estimated. Capital investment on the main route and for branch lines were:

	Main route	Branch lines
Piedmont route	46 million rubles	5 million rubles
Median route	34 million rubles	5 million rubles
Maritime route	31 million rubles	9.5 million rubles

Construction Times: Existing standards for gas pipeline construction did not vary greatly for the three options. However, experience indicated that the piedmont version would probably take much more time due to route-laying difficulties. The maritime route involved a greater risk of construction delays than the median route due to the need to cross marshy areas and three large rivers.

Gas Pipeline Maintenance: The piedmont route was recognized as the most difficult route to maintain, because access required the use of helicopters. The median version was the easiest to maintain as there was good access in all sections of the pipeline. The maritime route was intermediate in maintenance difficulty due to the presence of marshes.

The possibility of failure could not be ruled out in any of the three options. Experience with pipelines in other terrain conditions suggested that the maritime option would be the least reliable since a large section of the pipeline would run through an active corrosive medium (swamps). As the pipeline aged, the probability of failure would increase. In the piedmont option, experience of maintenance in mountainous regions included experience of landslides, which would be very difficult to completely avoid. The median option was the most reliable. Laying conditions were the most favorable, and it also had the best maintenance conditions.

Environmental Impact: The maritime route had the least adverse impact on the environment, due to passing through the marshy Kolkhida lowland area. The median route passed through unique and ancient forests, and passed through more agricultural land, citrus orchards, and tea plantations than did the other two options. Although these land losses would be temporary (only during construction), this type of environmental impact was still worse than it would be for the maritime option.

The piedmont option had the least attractive environmental impact. Cutting terraces into mountain slopes was a major problem, leading to increased probability of landslides. Terraces would also require greater amounts of land than would be needed by pipelines laid in level areas.

Connection to Regional Development Plans: The median and maritime routes would require roughly the same number of buildings to be demolished (69 buildings for the median option, 61 for the maritime option). The piedmont option was much worse at 136 buildings. The piedmont route also involved the most damage to agricultural cropland (129 hectares), followed by the median route (102 hectares) and the maritime route (57 hectares). Gas supply to potential customers was much better for the median route, so this was the option preferred by local authorities.

Construction Conditions: This criterion was greatly dependent on the relief and other physical characteristics of the district. The median route was assessed as the best option. The maritime route was rated as intermediate in preference, and the piedmont route by far the worst.

Population Safety: Standards for minimum distances from pipelines to residential areas were the primary means of obtaining population safety. In the event of a pipeline failure, a gas leak resulting in fire could occur. But regulations imposing working pressures and pipeline material standards made such a failure very unlikely. Of the three options, the maritime version was the most preferable with respect to population safety, since fewer settlements, agricultural land, and highways were exposed. The piedmont option and the median option were roughly the same with respect to population safety.

Overall Analysis

The criteria and expected alternative performance on each alternative are displayed in Table 1.

Table 1: Southern Pipeline Criteria and Performance

Alternatives					
	Criteria	Scale	Piedmont	Median	Maritime
C	Presented cost	million rubles	10.8	8.5	8.9
C1	Cost of laying route	million rubles	46	34	31
C2	Cost of branches	million rubles	5	5	9.5
Tmin	Construction time		worst	best	second
M	Convenience to maintain		inferior	best	inferior
R	Maintenance reliability		inferior	best	worst
IN	Environmental impact		far worst	inferior	best
RP	Regional development plans		worst	best	second
B	Construction conditions		far worst	best	second
S	Population safety		inferior	inferior	best

This data made it possible to eliminate the piedmont route. Further analysis was needed to select from the median and maritime options. Before reviewing that analysis, we consider the active groups involved in the decision.

Multiple Attributes, Multiple Actors

Four major participants were involved. The customer organization determined the design and was responsible for pipeline maintenance. There was a second organization that designed the pipeline. Regional authorities needed to approve any project, representing the interests of the local population. Finally, the contractor who would actually construct the pipeline was involved. These parties and the criteria they were most concerned with are given in Table 2.

Table 2: Parties and Interests in the Southern Pipeline Location Decision

Criteria		Project Organization	Authorities	Regional Customer	Contractor
C	Presented cost	x		x	
C1	Cost of laying route	x			
C2	Cost of branches	x	x		
Tmin	Construction time				x
M	Convenience to maintain			x	
R	Maintenance reliability	x	x	x	
IN	Environmental impact	x	x		
RP	Regional development		x		
B	Construction conditions				x
S	Safety	x	x	x	x

Each of these participants was concerned with a definite subset of the criteria given in Table 1 and in Table 2. The project organization was most interested in criteria C, C1, C2, IN, R, and S. Regional authorities were most concerned with RP, IN, S, R, and C2. The customer was most interested in C, M, R, and S. The contractor gave primary consideration to Tmin, B, and S. The sequence of activities in this decision are sketched in the following graph.

Sequence of Major Events: Southern USSR Pipeline

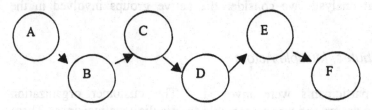

A	Need to move gas to consumers
B	Original two alternatives generated
C	Designer analysis of routes
D	Rejection of original alternatives
E	Creation of improved version
F	Decision

The Choice Process

The project organization analyzed possible routes from their perspective and made their choice. Then the project organization sent this proposal along with information about all three versions to the customer and then to regional authorities for approval. The contractor's representatives also participated in discussions. In this case, the project organization preferred the maritime version. Regional authorities pointed out the relative strengths of the two remaining options. The median version was far superior on criteria Tmin, C2, RP, and R and the maritime option was best on criteria IN and S. The contractor had to make tradeoffs between estimates for the median and maritime options. Thus it is apparent that the regional authorities, customer, and contractor all three faced difficult tradeoffs and were not satisfied with the existing alternatives.

Creation of a New Version and Final Choice

During the analysis the regional authorities asked the customer and the project organization to find new technical solutions to improve the evaluations of the median version on criteria S and IN in order to make them closer to these evaluations for the maritime version. This led to the project organization suggesting the possibility of reducing the guarding zone, combined with an increase in reliability from increasing the thickness of the pipe wall. Analysis determined that such an improvement would reduce the number of buildings requiring demolition considerably, and presented costs for the median and maritime options were very close despite the increase in the amount of metal required and the added cost of the pipeline.

New cost estimates for the median version were C = 9.5 million rubles, C1 = 40 million rubles, and C2 = 5 million rubles. Experts rated the new median version as approximately the same as the maritime version on criteria IN, and S. With these improvements, all participants in the selection process found the new median version the most acceptable, and this was the option selected.

CASE 2: Gas Pipeline Route Selection in Northern Siberia

Russia's current energy policy is to increase the proportion of natural gas in its national fuel balance. This policy is driven by an ecological preference for gas, as well as matching the more advanced technological capabilities of the Russian gas industry. Emphasizing the use of natural gas will support a strategic Russian industry and help stabilize the overall economy more quickly.

Over the past thirty years, the focus of Soviet governmental investment policy was exploration and development of new gas fields. Success in finding new fields increased the export of gas and encouraged further investment. Currently, the severe Russian federal budget deficit and the complex financial situation of the gas industry has resulted in more cautious consideration of new gas development projects. These analyses often use comprehensive systems analytic methods. They also are more open, explaining the decision process more thoroughly to Russian citizens (Osherenko, 1995)

The most promising prospect for additional gas is in western Siberia. This locale contains about 84 percent of the potential national gas reserves. Until now, only 44 percent of this potential has been accessed. The general decline in the country's industrial production has made the gas industry less prosperous. However, steady growth is expected in development of large new gas fields, primarily in the far northern regions of western Siberia.

The Yamal peninsula has become a key focus in new gas field development. Potential natural gas reserves in this region are estimated by various experts to be in the range of 10 to 20 trillion cubic meters. The Yamal gas and oil resources are recognized as one of Russia's great commercial assets.

The main element of a project to develop gas in the Yamal peninsula is a system to deliver gas from the west-central part of the peninsula to the primary pipeline in European Russia (Andre'eva, Larichev, Flanders, and Brown, 1995). This system will consist of six pipelines with a projected capacity of 146 billion cubic meters per year. These pipelines must cross Arctic and subarctic tundra with complex areas of permafrost and swamps dotted with numerous lakes. Building materials, such as gravel, are scarce. The length of the new pipeliine depends on the variant selected.

The locales where the gas pipeline route is to be located are for the most part unpopulated. Only small villages of local people are located in the Yamal Peninsula. But there are large herds of deer wandering over vast distances. The nature of the Arctic is very sensitive to human invasion. Unlike in the European part of Russia, natural damage is not self-restoring. The consequences of gas pipeline construction include tremendous destruction of Arctic land. Therefore, one of the main problems in the project is environmental damage.

Alternatives

After consideration of several variants, project planners selected two alternatives for detailed study. One option crosses Baidaratskaya Bay, a sea route. This pipeline would be 496 kilometers long. The second option stays on the Yamal Peninsula, turning west when it reaches the mainland, and crosses the Ural Mountains. This pipeline would be 654 kilometers long.

Each option has advantages and disadvantages. A feasibility study to clarify these options was prepared for the Russian gas company RAO Gazprom. RAO Gazprom employed two gas institutes for the study. The project institute in St. Petersburg preferred the land route. The other project institute advocated the sea route. There has been bitter discussion over several years between these two gas institutes over this location decision. The selection of the option was recently postponed, partly because of the complexity of the choice. The task is therefore one of selecting between the two options.

Criteria

Route length is an important criterion. The sea option would be 496 kilometers, and the land option 656 kilometers. But the sea option includes 68 kilometers of water crossing.

Construction conditions are difficult for both options. The land route would have to cross a large region of permafrost, rivers, and lakes. The sea route requires working in a limited period of 2 to 2.5 months each summer when the bay is free of ice. The portion of the sea route that crosses land has the same permafrost conditions and water obstacles as the land route, with the added complication of needing to cross about 68 kilometers of bay.

Construction cost is an important factor, but is very difficult to estimate. The land route cost can be based on the cost of labor, equipment, material, and transportation. Inflation of the ruble, however, makes any cost estimate difficult. The sea route's cost has additional uncertainties because there is little experience of construction in these conditions.

Construction time estimates are 5 to 7 years for both options. An unstable economic situation could impact the start of the project. This factor is of great importance, but difficult to estimate.

Environmental impact is important. Both options would negatively affect the environment. The land route would cross many ecologically important habitat areas and migration routes, including reindeer pastures, wildlife refuges, hunting lands, and rivers and lakes famous for their productivity and fishing resources. The sea route has less negative environmental impact because it does not cross valuable and vulnerable land on the Yamal Peninsula.

Risk of pipeline ruptures could be assessed based on data from other pipeline operations in the north of Russia. However, this risk on the sea route is more difficult to estimate. An underwater pipeline in severe Arctic conditions is unprecedented, so no accident statistics are available. Several features of the suggested route across Baidaratskaja Bay could induce accidents.

a) The shores of the bay are unstable because of nearly year-round permafrost action and the impact of sea ice.
b) Indications of ice scouring are present on the bottom of the bay.
c) Experts believe that iceberg sections in the Kara Sea are capable of reaching Baidaratskaya Bay.

Consequences of pipeline ruptures could be severe. An accident on the land-based pipeline would probably create an explosion and fire. This would completely destroy vegetative cover, upset the permafrost thermal regime, and possibly lead to the death of wild animals. Repair work would create further extensive and long-term environmental damage. An underwater accident would have much less environmental impact. Gas is soluble in water and is not toxic. The ice cover is not solid, and gas could escape through the cracks into the atmosphere. The appearance of an air inversion is not a concern in this unpopulated area and is unlikely.

Recovery from an accident would be quick for the land route, as pipe could be repaired immediately. Repairs in the underwater part of the sea route would be limited to the 8 to 10 weeks of time that the bay is free of ice. Furthermore, underwater repair would require special equipment and techniques.

Most of these factors involve unknown aspects. Long-term observation and prolonged investment in scientific research would be necessary for satisfactory characterization. Therefore, the site selection decision must be made under high degrees of uncertainty.

Evaluation of Alternatives by Criteria

The two project organizations prepared many reports, including calculations based on different models and assumptions. Their evaluations differ and sometimes contradict each other. This does not mean that insufficient information about the options was gathered. The reports include results of many years of observation conducted by special expeditions. Mathematical calculations based on models describing different Arctic natural phenomena are included. However, Arctic nature is too complex to be exactly captured in models. The only sources of information were experts. The manner information was elicited from these experts had to be carefully selected.

It is logical to consider only those criteria where the two options differ significantly. For example, the preliminary estimate showed that the construction time required for each option was roughly 5 to 7 years. Unstable national economic conditions could affect starting time, but this would be true for both options, and this factor can therefore be ignored because it is not relevant to the choice.

A second factor was that experts cannot be expected to accurately provide precise numerical estimates on these criteria due to the great degree of uncertainty and lack of objective data. Experts were, however, more than willing to provide comparative information about the alternatives based on their experience. The relevant criteria and their evaluations are given in Table 3.

Table 3: Criteria and Performances for Yamal Pipeline

Criteria	Alternatives	
	Land route	Sea route
Cost	C_{land}	C_{sea}*
Ecological impact	E_{land}*	E_{sea}
Accident probability	P_{land}	P_{sea}*
Accident consequences	Q_{land}*	Q_{sea}
Gas supply reliability	R_{land}	R_{sea}*
Uncertainty	U_{land}	U_{sea}*
* - indicates nonpreferred choice on that criterion		

Cost: The cost of crossing Baidaratskaya Bay (C_{sea}) was found to be a little higher than the cost of the land route (C_{land}). This evaluation is based on a study conducted by a foreign firm.

Ecological impact: Both options have negative impacts on the environment. Although the sea option includes some uncertainty, adverse impact is much larger for the land option, which crosses a lot of land and many rivers.

Accident probability: Because of unstable shores and the possibility of ice scouring, the probability of an accident is higher for the sea option.

Accident consequences: Accident on land usually involves an explosion and environmental destruction. The sea option would not involve an explosion, but gas would rise through the water and escape through cracks in the ice. The land option is clearly worse on this criterion.

Gas Supply Reliability: Repair of a pipeline after an accident requires much more time for the sea option, particularly since the bay is ice free for only 60 to 70 days per year. The sea option is clearly worse on this criterion.

Uncertainty: There are many uncertain and unknown factors connected with this project. The sea option clearly has more uncertainty and unknowns.

Multiple Attributes, Multiple Actors

Active Groups
Before final comparison of the two alternatives, we look at the decision process and its participants. Due to the high expected cost of the project, there will not be a single decision-maker. Several institutions and organizations are taking part in the decision process either directly or indirectly. These include five entities:

(1) The Russian joint stock company RAO Gazprom, which ordered the project development, and which must evaluate and confirm the pipeline route. RAO Gazprom's operational division in North Siberia is Nadymgazprom.
(2) The two project research institutes that developed the two options, each supporting their own project.
(3) The Ministry of the Economy, which evaluates the economic aspects of the project, and the Ministry for the Protection of the Environment and Natural Resources, which evaluates the ecological effects of the project.
(4) Local authorities in the Yamal region, who must give their agreement to the adopted option.
(5) Local communities and representatives of native peoples whose territory and resources will be affected by construction of the pipeline system.

These active groups have different interests, and it is expected that different groups will support different options. See Table 4.

Table 4: Parties Involved in Yamal Peninsula Pipeline Decision

Criteria	RAO Gazprom	Ministry of the Economy	Local Authority	Local Residents
Route length	x			
Construction conditions	x			
Cost	x	x		
Time	x	x		
Environmental impact			x	
Risk of rupture			x	x
Consequences of rupture			x	x
Recovery from accident	x	x	x	x
Uncertain factors	x	x	x	x

The sequence of activities involved with the Yamal Peninsula pipeline decision is sketched below:

Sequence of Major Events: Yamal Peninsula Pipeline

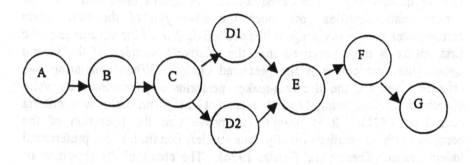

A late 1980s   Yamal project reopened as need to obtain more gas arose
B 1989   State expert panel questioned environmental
                     consequences of Yamal project
C 1993   Original two alternatives generated
D1   land version, Giprospetsgas
D2   sea version, Yusniiprograz
E 1993-1995   Gazprom held hearings about environmental issues
F   Ministry rejection of original alternatives
G   Creation of improved version

The Choice Process

The Verbal Decision Analysis approach was used to support the choice process. The framework of categorical or verbal decision analysis (Larichev, 1987, 1992) includes methods developed to appropriately compare alternatives on the basis of categorical evaluations over criteria. One of these methods is the pair-wise compensation method (PACOM, Larichev and Moshkovich, 1997).

First, the decision-maker ranks the disadvantages of the two options separately. In the Yamal pipeline case, disadvantages of the land and sea options were identified based on the decision-maker's preferences. Then reference options are constructed from real options. One reference option has the best characteristics on all criteria except for one, on which the alternative has a disadvantage relative to the other alternative. The other reference option is the same, but with a different criterion having the relative disadvantage. The decision-maker compares these two reference options, and identifies how much the advantage of the first option compensates for the advantage of the second option. If the advantage of the first option is more preferred than the relative advantage of the second option, the first option is the preferred choice. When comparing two reference options, the decision-maker performs a psychologically valid operation of comparing two objects that differ on only two criteria (Larichev, 1992). It is possible to prove that the operation of the compensation is mathematically true under conditions of preferential independence (Keeney and Raiffa, 1976). The check of the condition of preferential independence is the coincidence of results of comparisons for two pairs of reference options that differ only in the best or worst evaluations on all criteria except one or two when all other evaluations are at their worst state.

It is not always possible to find pair-wise compensation of disadvantages of two alternatives. When qualitative comparisons based on reliable decision-maker information are conducted, it is possible to have incomparable pairs of alternatives, when some evaluations are better for one option and others better for the second option. This is the case in the comparison of the land and sea alternative routes for the Yamal pipeline.

The following analysis is from the perspective of Gasprom. The sea option's greater uncertainty and lower reliability of gas supply are worse than the ecological disadvantage of the land option. But the negative consequences of an accident for the land option is worse than the larger probability of an accident for the sea option. In this case it is necessary to develop a new, more promising option on the basis of existing alternatives. The PACOM screen showing the tradeoffs between the land and sea options, with arrows indicating the preference relationship, is shown in Table 5:

Table 5: PACOM Screen of Tradeoffs

Criteria	Land Route	Baidaratskaya Route
1. Cost	C_{land} $\Rightarrow$	C_{sea}
2. Probability of an accident	P_{land} $\Leftarrow$	P_{sea}
3. Consequences of an accident	Q_{land}	Q_{sea}
4. Ecological impact	E_{land} $\Rightarrow$	E_{sea}
5. Reliability of gas supply	R_{land}	R_{sea}
6. Uncertain and unknown factors	U_{land} $\Rightarrow$	U_{sea}

The preference tradeoff input established that the combination of a lower attainment on probability of an accident paired with the top attainment for consequences of an accident was preferred to the reverse combination. The preference tradeoff input established that superior ecological impact combined with less reliable gas supply was inferior to the reverse case. The cost of the land option and the uncertain factors for the land option were superior. However, this information is insufficient to establish the superiority of either option. There is a lack of reliable information on which to compare the alternatives. The analysis conducted for other active groups demonstrated their disagreement.

Development of a New Option

In this case, as in many others, the practical value of decision analysis comes not only in comparing existing options, but in the creation of new options. ASTRIDA, a method for aiding strategic choice (Berkeley, et al., 1991) permits comparing several options, and also defining requirements for a new, potentially better option. ASTRIDA uses verbal evaluations of the options on each criterion, and calls for the decision-maker to conduct pair-wise comparisons. These comparisons can be conducted from the points of view of different active groups. Below we present an analysis reflecting the interests of RAO Gazprom.

In the case of incomparable alternatives, ASTRIDA leads to development of a new alternative through modifying an existing option. The method asks the question: What needs to be changed in one option to make it equal to or better than the other option? Thus, ASTRIDA defines the characteristics of a potentially better option that was not on the original list of preferred alternatives.

A new sea route option was identified from the search for ways to change the original sea options characteristics (Larichev, et al, 1995). Discussions with experts at the Yusniiprogaz Institute suggested ways in which negative aspects of the sea option could be removed.

(1) To eliminate the influence of seashore instability, special shafts could be constructed at a safe distance from the sea, and the pipeline placed in these shafts. This would incur additional construction costs, Cshafts.

(2) To avoid damage to the pipeline from ice scouring, the pipeline could be laid in special trenches 1.5 to 2 meters deep. These trenches would be deeper than original plans, so there would be a cost Ctrenches incurred.

(3) Icebergs are very rare but dangerous bay events. A special observation service with a ship capable of dragging icebergs away would eliminate this problem. This would involve an added cost Cice.

Adding these features to the old sea alternative generates a new option with an element of uncertainty approximately equal to that of the traditional land option. The probability of an accident for the new sea option is not very different than that of the land option. With the development of a special repair service for the underwater tubes, the reliability of the gas supply could be made equal to that of the land option. Thus, no significant differences now exist between the land and sea routes, except for cost and ecological impact. The cost of the new sea option is Csea + Cshaft + Ctrenches + Cice, clearly more expensive than the land option. The land option will still create greater environmental destruction. Now the comparison can be based on the differences in cost and environmental protection.

The analogical analysis was conducted from the positions of all active groups. The development of the new option was useful in this case. Based only on the original two options, local authorities and local populations were clearly in favor of the sea option. The project research institutes supported the options they each had developed. The new sea option was more attractive for the Ministry of Ecology and by Nadymgazprom. The positions of RAO Gazprom and the Ministry of Economy were finally influenced by the Russian financial situation, which has been rather difficult in the past few years.

While the final decision is pending, the methods described to identify the preferences for each group clearly brought focus to relative tradeoffs of the options, and led to improvement of one of the options.

CASE 3: Oil Pipeline Route Selection in Alaska, U.S.

The analysis of the two cases presented above was based on first-hand knowledge of projects as they developed. The third pipeline siting case is for a past decision that has been implemented. This presentation is based on published materials and discussions with some of the participants in the choice process.

A major recurring decision facing U.S. regulators charged with responsibility for Arctic development is whether or not to permit Alaskan oil construction projects to proceed, and if they are permitted to proceed, whether or not restrictions are to be imposed. There are several interested parties, including industry and environmental groups that have been vocally concerned with controversial decisions to include permitting oil drilling in the Alaskan National Wildlife Refuge. These decisions are often subject to arbitrary and unpredictable pressures, such as major shifts in the political climate. The interest groups have called for a standardized, scientific procedure that would not be susceptible to manipulation.

The U.S. Army Corps of Engineers Alaska District has lead responsibility for evaluation of permit applications using the wetlands guidelines of the Clean Water Act. Typically, one analyst within the Regulatory Branch makes a recommendation based upon available evidence, through the Branch Chief to the District Engineer, who in turn confers with various federal and state agencies before rendering a decision.

In the late 1980s, British Petroleum sought permission to develop the Niakuk oil field from an artificial island 1.25 miles off the Arctic Beaufort Sea shore. A gravel causeway was planned to pipe the oil ashore. The Alaska District of the Corps of Engineers gave a conditional permit that did not allow for the construction of the proposed causeway on the grounds that the proposal did not meet an agreement between the oil companies and federal regulators to maintain fish habitat. A regulatory guideline stated that irrespective of other considerations, a permit has to be denied if a fish population was adversely affected. The agreement among the companies and agencies had been to judge potential changes in fish populations based upon changes in habitat. Causeways were believed to affect the habitat of Arctic cisco by changing coastal current patterns. The experience accumulated after the construction of West Dock and Endicott causeways gave evidence of negative influence on fish habitat. Although the senior official judged the causeway to have overall social value, under existing regulatory procedures regulators were not allowed to trade off some negative environmental impacts against other positive values.

The Alternatives

According to the existing rules, the Alaska District Corps of Engineers could not simply reject the proposal. The regulator also had to show an alternative option to the applicant. The Alaska District prepared a document presenting British Petroleum with two options:

(1) Slant drilling from the shore
(2) A subsea pipeline.

These two options were stated to be environmentally acceptable. The Alaska District recommended that British Petroleum adopt the slant drilling option as the less expensive alternative.

Multiple Attributes, Multiple Actors

There were three primary active groups that participated in the decision.

(a) The public interest was represented by the Alaska District of the Corps of Engineers with respect to technical matters, and the Environmental Protection Agency with respect to environmental matters

(b) Oil companies, specifically British Petroleum

(c) Local authorities, who had final approval for the project.

Criteria and Evaluation

This case has been analyzed by two different multicriteria methods: multiattribute utility theory and verbal decision analysis (see details in Flanders, et al., 1998). We will outline analysis of this case using the approach of verbal decision analysis (Larichev and Moshkovich, 1997; Flanders, et al., 1998). In this approach, only the most important and evident differences between the alternatives are considered. Generally speaking, it is possible to find many criteria for evaluating alternatives. But verbal decision analysis seeks to use only the "broad brush strokes" in the uncertain "picture" of information available, presenting the essential differences to the decision-maker and active groups.

Verbal decision analysis involves a five-step process:

1) Elicitation of problem descriptions acceptable to the applicant and to the regulatory bodies

2) Description of alternatives by both the applicant and the regulatory body in authority

3) Comparative evaluation of alternatives in ordinal form

4) Structuring evaluation of alternatives by the regulatory body with possible alternative rejection

5) Development of a psychologically valid prescriptive procedure of choice acceptable to both the applicant and to the regulatory body.

Table 6 shows a qualitative analysis of the Niakuk decision. The analysis, which uses seven criteria, was conducted from the perspective of a regulator.

Table 6: VDA Hypothetical Qualitative Analysis of Niakuk Project

Active Group	Criteria	causeway	slant drilling	subsea pipeline
		Alternatives		
ABC	Impact on the Environment	Unacceptable	*Acceptable	*Acceptable
AB	Damage to the	Severe	*None	Light
ABC	Quantity of oil	*Basic volume	Less	*Equal to basic
C	Cost of Construction	*Basic cost	$14-$30 million (5 to 10% more)	$37 million (10-15% more)
A	Oil independence	*Basic estimate	Less	*Equal to basic
ABC	Uncertain factors	*A few	Very many	Many
ABC	Reliability of Pipeline	*Normal	*Normal	Less

Active groups: A-State of Alaska
                 B-Corps of Engineers
                 C-British Petroleum
*-indicates preferred attainment level

This list of criteria in Table 6 is taken from Flanders, et al. (1998). Of the primary parties, the State of Alaska represented local groups, the Corps of Engineers had governmental regulatory responsibility, and British Petroleum proposed the project. The Environmental Protection Agency also represents citizen concerns related to protecting the environment.

The Choice Process

For permitting decisions of this nature, it is typical to have two primary active parties. Regulatory organizations like the Corps of Engineers or the Environmental Protection Agency represent the government's view of public interest. The applicant with supporting organizations such as the State of Alaska and other sponsors typically present a pro-project view. The general sequence of events in this decision are sketched below.

Sequence of Major Events: Niakuk, Alaska

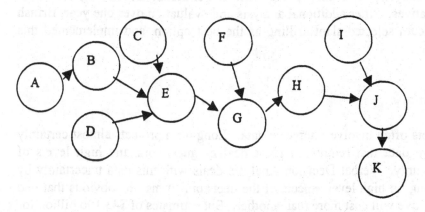

A	1968	Prudhoe Bay oilfield discovered
B	1971	Alaska Native Claims Settlement Act made pipeline construction possible
C	1974	U.S. concerns about oil availability
D	1980s	increased environmental concerns
E	late 1980s	British Petroleum presented proposal with cost, benefit, environmental evaluations
F		Hearings
G		Corps of Engineers rejected original proposal on basis of environmental risk, generated two alternative proposals
H		British Petroleum originally rejected alternative proposals
I	1991	Congress voted against development in Arctic National Wildlife Refuge
J		British Petroleum accepted slant drilling alternative
K		British Petroleum operating successfully

Because of the threshold for anadromous fish, which was a component of the environmental impact criterion, the causeway proposed by British Petroleum was deemed unacceptable and was eliminated as an option. British Petroleum then faced the choice between the remaining two options. The major differences between the two remaining options were found to be the cost of construction, the number of uncertain factors, and pipeline reliability. The subsea pipeline had relative disadvantages in cost and reliability.

We have no knowledge of how the British Petroleum staff conducted their analysis. It is known that British Petroleum initially rejected both alternatives. After additional analysis and evaluation over one year, British Petroleum selected slant drilling as the best option, and implemented this decision.

Summary

Projects often involve imprecise data. Long-term projects almost certainly involve data that requires a great deal of guesswork and high levels of uncertainty. Verbal Decision Analysis deals with this data uncertainty by focusing on high level aspects of the decision. It may be obvious that one alternative will cost more than another. But estimates of $43.106 billion for one and $82.545 billion for the other are both highly unlikely to be the actual outcome. The decision usually depends on the relative degree of cost. Verbal Decision Analysis allows avoiding the great effort required to obtain detailed and inaccurate cost estimates and replaces them with their expected relative cost, or other criterion measurement. This allows decision-makers to isolate relative advantages, and to conduct a psychologically valid decision analysis.

The method also allows development of new, superior alternatives. This was demonstrated in all three cases given in this chapter. The means to accomplish superior alternative development is to focus on the weaknesses of a particular alternative, and to improve them, identifying rough costs for these improvements so that comparisons of alternatives can be continued.

Verbal Decision Analysis provides a promising, sound method for high-level decision-making when imprecise data is present. It requires far less detailed calculation, and thus requires much less decision-making time, than MAUT and outranking methods. Since its approach is generally what top-level decision-makers will use to combine criteria on different scales in the final phases of these other analyses, it seems very promising as an overall method.

References

Andre'eva, Y., Larichev, O., Flanders, N. and Brown, R. Complexity and uncertainty in Arctic resource decision: The example of the Yamal pipeline. *Polar Geography and Geology* 19, 1995, 22-35.

Berkeley, D., Humphreys, P., Larichev, O., and Moshkovich, H. Aiding strategic decision making: Derivation and development of ASTRIDA. In Y. Vecsenyi and H. Sol, eds., *Environment for Supporting Decision Processes*, North-Holland, Amsterdam, 1991.

Dey, P.K., Ogunlana, S.O., Gupta, S.S. and Tabucanon, M.T. A risk-based maintenance model for cross-country pipelines. *Cost Engineering* 40:4, 1998, 24-31.

Flanders, N.E., Brown, R.V., Andre-eva, Y. and Larichev, O. Justifying public decisions in Arctic oil and gas development: American and Russian approaches. *Arctic* 51:3, 1998, 262-279.

Goncharov, V.I. and Oseredko, Y.S. *Questions of systems approach to gas pipeline systems designing with the help of computers*. Kiev, Znanie Society of the Ukrainian SSR, 1977 (in Russian).

Keeney, R.L. and Raiffa, H. *Decisions with Multiple Objectives: Preferences and Value Tradeoffs*, John Wiley & Sons, New York, 1976.

Larichev, O. *Objective Models and Subjective Decisions*, Nauka Publishing House, Moscow, 1987 (in Russian)

Larichev, O. Cognitive validity in design of decision-aiding techniques. *Journal of Multi-Criteria Decision Analysis* 1:3, 1992, 127-138.

Larichev, O., Brown, R., Andreyeva, E. and Flanders, N. Categorical decision analysis for environmental management: A Siberian gas distributing case. In J.-P. Caverni, M. Bar-Hillel, F.H. Barron and H. Jungermann, eds., *Contribution to Decision Making*, North-Holland, Amsterdam, 1995, 255-286.

Larichev, O. and Moshkovich, H. *Verbal Decision Analysis for Unstructured Problems*, Kluwer Academic Publishers, Boston, 1997.

Oseredko, Y.S., Larichev, O.I., and Mechitov, A.I. Main gas pipeline route selection problems, taking into consideration risk and uncertainty factors. In H.C. Kunreuther and E.V. Ley, eds., *The Risk Analysis Controversy*, Springer-Verlag, Berlin, 1982, 91-101.

Osherenko, G. Indigenous political and property rights and economic/environment reform in northwest Siberia, *Post-Soviet Geography* 36:4, 1995, 225-237.

8 PROBLEMS AND TOOLS

People face multiple criteria in many decision contexts. Business people quite often face the problem of siting a location for various elements of their enterprise. Nuclear physicists need to select sites for the disposal of nuclear waste. These problems are complex, just as that of an individual looking for the appropriate drug to provide needed vitamins. There are many alternative drugs available, each with their own potential for complications. The advice of a physician is needed in this selection problem. Similarly, specialists in decision making with strong practical experience are extremely useful in enabling people to deal with complex multicriteria problems.

In this chapter we will offer some guidelines for people to apply multicriteria analysis to help make location selection decisions. We will describe the particular features of multicriteria location decision problems, and present brief versions of the different decision theory paradigms to enable identifying the appropriate multicriteria approach for a specific decision.

Specific Features of Multicriteria Location Selection as Decision Problems

Let us review the specific features of multicriteria location problems (MLP) as a decision problem.

1. Usually each multicriteria location problem is unique: either a new decision, or a repetitive decision under very different conditions from what has been experienced previously.

2. Multicriteria location problems involve strategic choice: solution is needed a long time before it can be implemented.

3. The problem is to select one best alternative from a group of alternatives.

4. The group of alternatives is usually given. In some cases when a decision-maker and/or active groups are not satisfied by the given set of alternatives (before or after the choice) an additional alternative could be provided by experts. But after this stage, there is again the given set of alternatives for the estimation and selection of the best.

5. The number of criteria is different, ranging from 2-3 in simple cases to several dozen in difficult problems (see Chapters 4 and 6).

6. The level of uncertainty in the evaluation of alternatives could be also vary, ranging from very low (the cost of restoring a waste disposal site to its previous state) to very uncertain (the probability of severe accident or the number of years before a catastrophe might expose materials stored in the repository).

7. There are several active groups involved in the choice process (as shown in Chapters 3 through 7). In the simple case a specific decision-maker would have choice discretion. But in the general case the choice of the best alternative must be accomplished through the process of negotiations between the decision-maker responsible and active groups.

The Different Paradigms in Decision Making.

In Chapter 2 four different multicriteria approaches were presented based on different paradigms.

1. In English-speaking countries, the emphasis has been on implementation of the idea of utility, where value is expressed as some function of the preference importance of for each criterion times the performance of the given alternative on that criterion (Keeney and Raiffa, 1976). This argument is based on the concept that there is a utility (or value) function, and the alternative with the greatest utility is the choice that the decision-maker should prefer. There have been a number of techniques that have been developed to implement this idea. The best-known are MAUT-multiattribute utility theory (Keeney and Raiffa, 1976, Dyer and Sarin, 1979) and SMART (Edwards and Barron, 1994). The particular features of this paradigm are:

a) The utility function is constructed without taking into account the number of alternatives. It means that the same decision-maker's efforts are the same for a decision with 3 alternatives as for one with 50 alternatives.

b) There is the presupposition that the estimations on all criteria could be expressed quantitatively, and that the transformation of qualitative ideas into numbers is always reliable.

2. The analytic hierarchy process (Saaty, 1980) uses a different approach to estimate relative value of criteria as well as scores of alternatives over these criteria. The pair-wise comparison of criteria (by importance) and alternatives (on criteria) is the special feature of this approach. There is also the presupposition that the estimations on all criteria could be expressed quantitatively, and that the transformation of qualitative ideas into numbers is always reliable. A universal scale of such a transformation is used.

A number of methods have branched off from AHP, including the geometric mean approach (Barzilai, et al., 1990), and REMBRANDT, the basic ideas of which were outlined in Lootsma (1993). HIPRE 3+ (Hämäläinen, 1992) is a software allowing the decision-maker to implement AHP or SMART modes of input (as does REMBRANDT). Interval bounds for AHP are considered in preference programming (Salo and Hämäläinen, 1995).

3. The outranking methods coming from France and Belgium use a rather different paradigm. The alternatives are compared by pairs. Based on the concepts of B. Roy (1968), alternatives are compared on the basis of two measures: concordance and discordance. Concordance reflects those cases where one of a pair of alternatives is superior to the other. Discordance reflects the reverse case, where one of a pair of alternatives is inferior to the other. Another concept involved in outranking is incomparability (Roy, 1996). When the index of concordance is low and index of discordance is high, the alternatives in a pair are incomparable. The best known implementation of this system of ideas is the family of ELECTRE outranking relationships based on weighted concordance and discordance indices. PROMETHEE (Brans and Vincke, 1985) is another popular implementation of these ideas.

4. The approach of Verbal Decision Analysis is based on different paradigm. The major emphasis is given to the possibilities and limitations of human information processing system. The process of information elicitation from the decision-maker is based on psychologically admissible questions (Larichev, 1992). The example of such a question is the comparison of two evaluations from the scales of two criteria (see Chapter 7). The information provided by the decision-maker is checked for possible contradictions. There is also the concept of incomparability but in a different sense. The alternatives are incomparable when they could not be compared with the help of psychologically valid operations of information elicitation. Different VDA methods have been used for the solution of different decision problems (Larichev , Moshkovich, 1997). The method PACOM is used for the selection of the single best alternative from a group of alternatives (Chapter 7).

Comparison of Methods in Laboratory Experiments

Over the past few years numerous laboratory experiments comparing multicriteria methods comparison have been conducted. One is presented below, demonstrating clear difference between MAUT and VDA approaches in the problem of selection of best from several alternatives. The experimental study compared verbal and numerical methods of decision making (Larichev et al., 1995). The subjects were college students nearing graduation, who were in a job search process, facing opportunities similar to those given in the study.

Let us suppose that a college graduate has several offers (after interviews) and he (or she) is to make a decision. Every variant is acceptable, but of course, one variant is better upon one aspect but inferior on others. So, the student has a multicriteria problem. The student was asked to solve this decision problem with the help of the appropriate multicriteria method.

Let there be Q criteria, upon which N alternatives are evaluated. Each alternative a_i (i=1,2, ..., N) corresponds to the vector $a_i = (a_{i1}, a_{i2}, ..a_{iQ})$.

Four criteria are used as the focus for the study: *salary, job location, job position* (type of work involved), and *prospects* (career development and promotion opportunities). The following alternatives were used :

FIRM	SALARY	JOB LOCATION	POSITION	PROSPECTS
a1	$30 000	Very attractive	Good enough	Moderate
a2	$35 000	Unattractive	Almost ideal	Moderate
a3	$40 000	Adequate	Good enough	Almost none
a4	$35 000	Adequate	Not appropriate	Good
a5	$40 000	Unattractive	Good enough	Moderate

In this case there are three possible values upon each criterion. The greater the salary, the more attractive it would be to a rational subject. Thus, we have four criteria with three possible values each and the values upon each criterion are rank-ordered from the most to the least preferable.

It is evident that there are no dominated alternatives. Therefore, comparison of these alternatives requires some value function, which would take into account the advantages and disadvantages of each alternative upon each criterion.

Two decision support systems based on Multiattribute Utility Theory (MAUT-Keeney and Raiffa, 1976; Keeney,1992) were used for the solution of the problem given above. These systems are LOGICAL DECISION (Smith and Speiser,1991) and DECAID (Pitz,1987). The third DSS was one based on Verbal Decision Analysis (see below).

Both LOGICAL DECISION and DECAID decision support systems implement ideas of multiattribute utility theory, providing the ability to construct an additive utility function for the case of risky decisions, and an additive value function for decision making under certainty. The value function obtained from both systems would therefore have the linear form,

$$v(a) = \sum_{i=1}^{Q} k_i v_i(a_i)$$

where a is an alternative estimated over each of the Q criteria, k_i is the coefficient of importance for the i-th criterion , a_i is the value of alternative a on criterion i , and v_i is the value function for the i-th criterion.

Both systems are easy to use, have flexible dialogue and graphical tools to elicit the decision-maker's preferences. The main difference in the systems (besides their interfaces) is the way in which numerical values are determined upon separate criteria and the determination of criteria weights. In DECAID pure graphical (direct) estimation is used (a point is placed on the line ranging from 0 to 1). In LOGICAL DECISION, criterion values are identified in one option by the decision-maker marking the "middle" value for the criterion to determine the parameters of this function.

Criteria weights are also defined in a different manner in these two systems. In LOGICAL DECISION criteria weights are defined on the basis of trade-offs in a traditional manner (Keeney, 1992). In DECAID weights are elicited directly (graphically – by selecting a point on a line). DECAID also provides the opportunity to make trade-offs, but after that operation the result is presented as points on lines. Thus, it is possible to consider DECAID as using direct elicitation of criteria weights.

Considering the common features implemented in both systems and also the similarity of information received from a decision-maker in the process of task solution, the attempt to solve the above described task with the help of these systems should lead to the same result.

The third DSS is one from the family of Verbal Decision Analysis (ZAPROS - Larichev and Moshkovich, 1997). Only verbal measurements are used on all stages of this method. ZAPROS uses ranking rather than rating information, but the additive overall value rule is correct if there is an additive value function. In ZAPROS the additive rule does not provide the summation of values, but rather the means of obtaining pair-wise compensation between components of two alternatives.

For preference elicitation from subjects the following procedure was used:

Subjects were asked to compare several pairs of alternatives formed to trade-off levels of two criteria. For each pair two alternatives were different only on two criteria evaluations (one evaluation was best for each alternative) and had equal evaluations (best or worst) on other criteria. In the specific task used subjects compared pairs of alternatives different on each of the four criteria. An example of a typical question is: "Which do you prefer: a salary of $40,000 with an adequate location, or a salary of $35,000 with a very attractive location? Please take into account that on the criteria «Position» and «Prospects» both options have ratings of «good»."

Comparing these alternatives, subjects were to choose one of the following responses:

1. alternative 1 is more preferable than alternative 2;

2. alternative 2 is more preferable than alternative 1;

3. alternatives 1 and 2 are equally preferable.

Implementation of this simple system for comparison of pairs of alternatives allows a simple check of the comparisons provided with respect to transitivity:

1) if $a > b$ and $b > c$, then $a > c$;
2) if $a > b$ and $b = c$, then $a > c$;
3) if $a = b$ and $b = c$, then $a = c$;

where a, b, c are alternatives, symbol $>$ means more preferable and symbol $= $ - equally preferable.

The method thus provides verification of the comparisons provided for transitivity and allows the subject to change some responses to eliminate intransitivity. It also guarantees that comparison of each pair of alternatives from this set is supported by at least two responses provided by the user.

This method of preference elicitation is psychologically valid (Larichev,1992). The information provided by the decision-maker allows building a joint ordinal scale combining all evaluations on separate criteria scales. The joint ordinal scale is used for partial ranking of alternatives, and is applied to the specific alternatives provided to the system. This rank-ordering may be used for comparison of the initial 5 job opportunities because in our task the additive value function is appropriate and criteria were formed to be preferentially independent. This algorithm does not guarantee comparison of all alternatives because for some pairs of alternatives ZAPROS gives only an incomparability relation.

Each subject from the group used all three systems for the solution of the problem presented above. The difference in the outputs of methods consisted in following: some pairs of alternatives had not been compared with the ZAPROS method. The method of preference elicitation used by ZAPROS does not in general provide a basis for ranking all possible alternatives (only a partial ranking is obtained). LOGICAL DECISION and DECAID gave numerical values of the utility for all alternatives, which were used for complete ranking.

The results of the experiment were analyzed in different forms: the ranking of given alternatives, the ranking of special alternatives used in ZAPROS, the ranking of criteria weights and so on.

A very low correlation was found in the outputs of LOGICAL DECISION and DECAID. The ANOVA test demonstrated that for the group of subjects the outputs of LOGICAL DECISION and DECAID were not significantly different in measurements of criteria weights and ranking of alternatives.

The following results were very interesting: the outputs of pairs LOGICAL DECISION-ZAPROS and DECAID- ZAPROS were correlated at a statistically significant level. This implies that only for alternatives compared by ZAPROS were the relations essentially the same.

The following explanations are possible:

The alternatives that could be ordered by ZAPROS are closer to ordinal dominance. Ordinal relations are more stable than cardinal relations. Moreover, the ZAPROS relationships were constructed in a very reliable manner using verbal measurements, a psychologically correct means of preference elicitation with transitivity checks of information provided and elimination of contradictions.

Two complete orders were constructed by LOGICAL DECISION and DECAID based on cardinal measurements and the weighted sum of alternative utility estimates by criteria. The difference in the utility (no matter how small) defined the final order of alternatives. The errors (even when small) made by subjects in providing numerical measures resulted in quite different ordering of alternatives.

Comparison on Practical Tasks

We know of a small number of studies comparing different multicriteria approaches on practical tasks. Such comparisons are very important. Laboratory experiments can shed some light on methodological problems (see above) and can be very informative. But student behavior in experiments is expected to be different from that of mature decision-makers. On the other hand, it is impossible to organize experiments with mature decision-makers: they do not play games in real world situations.

Therefore, some different approaches are needed. We shall present below several attempts to organize some comparisons on practical tasks involving multiple criteria.

MAUT and Outranking Approaches

Roy and Bouyssou (1986) systematically compared MAUT and outranking (represented by ELECTRE III method) on the practical case of siting a nuclear power plant. The methodology was to apply the outranking approach to the case described in the paper of Keeney and Nair (1976). The goal of this comparison was to demonstrate the differences between the approaches on a practical case. The task consisted in selecting the best of nine alternative locations evaluated in the terms of six criteria. The comparison demonstrated that: 1) the same task could be solved by different methods; 2) the difference between the methods is large.

Roy and Bouyssou stressed the following criteria for evaluation of the methods:

1. Foundation:
MAUT has a strong mathematical foundation. The form of utility function is taken on the basis of testing axioms of utility theory. The justification for the outranking approach is the "application of common sense" to the principles of avoiding the situation when "a major disadvantage on one criterion could be compensated for by a large number of minor advantages on other criteria" (Roy and Bouyssou, 1986, p.201) in pair-wise alternatives comparison.

2. The perception of decision-maker preferences.
In MAUT, the analyst collaborates with the decision-maker to elicit the complete system of pre-existing preferences. Sensitivity analysis could be applied to check for the influence of small changes in the preferences on final output of the analysis.

The goal of the outranking approach is "trying to design a preference system" for the decision-maker, to give information about pair-wise comparison of alternatives by different levels of concordance and discordance indices. This means that the decision-maker has some time to develop the preference system in the interactive process.

3. The procedure of preference elicitation.

In MAUT, the well known lottery technique (Raiffa, 1968) is used to evaluate the parameters of the model. Roy and Bouyssou (1986) stressed that the questions to the decision-maker are "extremely complex" and the decision-maker "may be unable to reply to such questions in a reliable fashion."

In the outranking approach, the decision-maker is asked to nominate the weights, and the veto threshold on each criterion. Those parameters are "designed to convey deliberate positions adapted by the decision-maker, positions which are mainly of a qualitative nature." (Roy and Bouyssou, 1986, p.208).

4. Measurement of estimates on criteria.

In the MAUT approach, all parameters are expressed in a precise quantitative way. In the outranking approach, it is not important to know a probability distribution "with a possibly misleading precision" but to find the essential difference between the alternatives (Roy and Bouyssou, 1986, p.212).

5. The output of the analysis.

In MAUT, the complete ranking of nine alternatives by utility is given.
In the outranking approach, two partial orders are given (in the dependence from the form of one from the criteria) including an incomparability relation between some alternatives (Roy and Bouyssou, 1986, p.211).

Comparison of MAUT and VDA

Larichev et al. (1995) and Flanders et al. (1998) present the results of the comparison of MAUT and VDA on practical tasks. The idea of comparison suggested by Professor R. Brown was to use both approaches on similar practical tasks (resource allocation on Arctic projects) in the US and in Russia. The consultant with practical experience used his own approach for solving the problem. The representative of the other scientific school had the opportunity to demonstrate the application of different approach during the meeting with the decision-maker. The practical cases used are presented in Chapter 7: the choice of gas pipeline route on Yamal Peninsula and the choice of oil pipeline route on Niakuk island. MAUT was implemented using the graphical method of information elicitation (Flanders, et al., 1998).

After the application of the two quite different approaches, MAUT and VDA were compared on the general pros and cons of each approach, and the circumstances under which one or the other would be favored. It is clear that MAUT is much more widely known and has to date been applied many more times than VDA. But the potential of development for every approach is to be connected with investigation of possible limits of its applicability in practical tasks. In our opinion, three groups of criteria for comparison are to be taken into consideration: methodological, institutional and personal (Flanders et al., 1998).

Methodological criteria: Methodological criteria characterize an approach from following perspectives:

> how the measurements of alternatives on criteria are made;
> consideration of alternatives
> how psychologically sound people are in making such measurements
> (complexity reduction);
> how a desirable decision output can be obtained;
> mitigating uncertainty
> the cognitive burden and time required by an approach;
> how satisfactorily are major problems with informal analysis
addressed?

Measurement of option evaluations on the criteria. There are several reasons for people to give and receive an information in verbal form:

1. People use verbal communication much more easily than quantitative communication. Probability theory appeared many thousands of years after development of languages.
2. Words are perceived as more flexible and less precise, with various communicative functions, and therefore seem better suited to describe vague opinions and characterize imprecise beliefs.

Erev and Cohen (1990) stated that " forcing people to give numerical expressions for vague situations where they can only distinguish between a few levels of probability may result in misleading assessments".
But there are positive factors in utilization of quantitative information:

1. People attach a degree of precision, authority and confidence to numerical statements that they do not ordinarily associate with verbal statements.
2. It is possible to use quantitative methods of information processing (e.g. Bayes theorem).

The experiments made over many years by Prof. T. Wallsten and his colleagues demonstrated no essential differences in the profit received by subjects or in the accuracy of evaluations.(see e.g. Budescu and Wallsten, 1995; Erev and Cohen, 1990).

The most salient difference consisted in the number of preference reversals (Gonzalez-Vallejo and Wallsten,1992) between binary choice decisions and the bids for lotteries based on events. Preference reversal is quite common in human behavior. The frequency of predicted reversal was significantly decreased in the verbal mode relative to the numerical display mode. This means that subjects were much more consistent when using verbal expressions instead of numerical expressions.

Huber and Huber (1987) studied comparative verbal probabilities in a systematic way. We emphasize the following results. Lay-people (adult as well as children) use comparative verbal probabilities much more often than numerical expressions of probability. Comparative probabilities are used for tasks of an objective nature (spinning a circular disc) as well as for tasks of a subjective nature (sport competitions, games). According to Huber and Huber, the evaluation of comparative, verbal probabilities are much more reliable than for quantitative probabilities.

Consideration of Alternatives: The two methods differ considerably in whether they force consideration of alternatives. The qualitative approach seeks resolution of the decision problem by engendering a search for another alternative that has not previously been considered. As seen in the Yamal case, this alternative can be a new option developed by altering the initial negative features of one option. The quantitative approach (although it can consider other alternatives) does not use option creation as a tool in the analysis. Someone looking for alternatives could use the qualitative analysis to see where the largest disadvantages of the existing options are and try to develop a third option based on these observations. Here again, the quantitative approach does not force considerations of alternatives, but reduces differences to numbers.

Complexity reduction: The qualitative approach eliminates much of the complexity by reducing questions to their bare essentials due to differences. In the Yamal pipeline case, this difference came down to a trade-off between environmental impacts and cost. The qualitative approach can reduce several factors down to numbers, as was the case in reducing uncertainty down to a dollar value, but it does not try to equate everything to dollars. This approach has the advantage of clearly emphasizing main differences.

The quantitative method does not so much reduce the complexity per se, but brings to bear a common denominator. A chief issue between the two methods is whether the numbers created under the quantitative approach are "real:" do they truly reflect psychological states, that is states in which numerical intervals are equidistant and values placed on different impacts in fact follow a common scale and are thus comparable?

Providing desirable decision output: MAUT gives the value of the utility of each alternative. This makes it possible to not only identify the best alternative but also to define the difference in utility between alternatives. This means that the output of MAUT methods is rich enough to give the decision-maker the basis for detailed evaluation and comparison of any alternatives.

By using the method PACOM based on the VDA approach one is trying to construct a binary relation between the alternatives. Two alternatives (options) could have relations of dominance, equivalence and incomparability. In the last case they could not be compared on the basis of psychologically valid operations of information elicitation. This means that the VDA approach can not give recommendations to the decision-maker if the alternatives to be compared are incomparable. *Mitigating uncertainty.*

A goal of all decision methods is reducing the confusing effect of uncertainty. The approaches deal with uncertainty in very different ways. VDA, at least in the Yamal case, looks at the cost of reducing uncertainty. That is, it alters uncertainty into a monetary figure: how much will reducing uncertainty to an acceptable amount cost? Even if decision-makers do not decide to pay for that reduction, a precise value of uncertainty is given. A public debate can be pursued that would yield a precise figure. The MAUT approach attempts to estimate the amount of uncertainty. Here again the decision-maker has to be able to think in numerical terms to provide an estimate. The payoff is that the analysis can derive a single estimate of uncertainty to go with the single estimate of utility. Again, the uncertainty can be graphically presented

Time spent and cognitive burden on the decision-maker. From the practical point of view it is important to stress the features of the method that minimizes time needed and cognitive burden. We have examples for both approaches. In the case of MAUT, the elicitation of decision-maker preferences for a real case similar to Niakuk involving two options with 30 criteria obtained from a project manager with a baccalaureate degree in biology who was unfamiliar with the approach took about 4 hours. She had been concerned about diverting time from a time-pressure situation in order to participate in the exercise. But afterwards she said that the process of elicitation, with immediate computer feedback about the implications, helped her to do analysis she would have had to do anyway relative to recommending a decision on a permit request. She felt that MAUT improved her thinking. She did not find the answering to be cognitively uncomfortable.

In the Yamal case the decision-maker spent less than one hour answering the questions needed to compare two initial options. So the time burden was less than that of MAUT. The questions were posed in natural language and the decision-maker really did not suspect that a decision method was being used.

Institutional criteria

Institutional criteria include: how communications in and between organizations could be improved (the ease of using the approach); and consequences of cultural differences.

Improvement of communication: Both approaches may be considered improvements over the often-confounded discussions surrounding oil and gas development in the Arctic. Achieving greater clarity does, to some extent, provide improved communication.

The VDA approach uses more direct communication because sentences are taken from the language used by the decision-maker and active groups are used to compose the verbal quality grades upon the criteria scales. The VDA approach does not require the decision-maker or expert to have previous knowledge in decision methods. The method is "invisible " to a user. A method should be a completely natural tool for a user, adapted to the way in which information is exchanged within an organization and between organizations.

Conversely, MAUT findings can be presented graphically because of its numerical base. The research did not specifically attempt to measure improvements in communication, but conversations with regulators who work with the Corps of Engineers indicated that they agreed that the analyses were improvements. This area will be the subject of future research.

Cultural differences: It is necessary to stress the cultural differences influencing the applicability of different approaches. Americans use numerical evaluations more typically than do Russians. Acceptability of numerical evaluations even in the cases of difficult public choice is demonstrated by contingent valuations in terms of dollars. This requires analysts "to put a price tag on goods not traded in any market place " (Fischhoff ,1996). The requirements to apply cost-benefit analysis for all significant federal actions have defined special conditions of analytical work for American analysts. Thus numerical decision analysis is accepted more widely in the US than in Europe.

Personal criteria

Personal criteria include: educational level required of decision-makers to use methods; and how consultant professional habits influence the selection of an approach.

Educational level of a decision-maker: Decision-makers need to know how to obtain useful output from the analysis and to know how to use a consultant's help. It is a question of practical experience and intellectual ability on the part of the decision-maker. Such qualities are presuppositions for the utilization of any analytical technique.

Personal education in decision analysis helps decision-makers to understand and accept the MAUT approach. VDA methods do not require any special knowledge in decision analysis on the part of the decision-maker. The VDA approach is especially needed in the situations when a decision is made under new circumstances or in conditions of uncertainty.

Professional habits of a consultant: The consultants working in different countries often use quite different analytic techniques in the process of analysis. It is evident that French consultants use ELECTRE methods (Roy,1985) much more often than AHP methods (Saaty,1980). VDA methods are used by Russian consultants. MAUT and AHP are the most popular decision analysis methods in the US.

This situation can be partly explained by the correspondence of an approach to the culture of decision making in the country. For example, in the US monetary evaluations of nonmonetary goods is often required.

Common Stage for All Approaches: Systems Analysis.

The application of a decision method in any practical task is the combination of art and science (Larichev, 1979) - the combination of some technique and skill of a consultant. Therefore, any of the four multicriteria approaches presented above use a systems approach to a problem. This means that the preliminary conditions for the utilization of a technique are:

-selection of the set of criteria;
-collection of the set of initial alternatives;
-clear definition of active groups;

Those steps require consultant effort and skill. In any practical case the success of the overall analysis requires success during this stage.

The next stage consists in the application of a decision method itself. In real life it is difficult to differentiate stages because they are closely connected: the second stage of the analysis requires measurement of factors following sound procedures, involving active groups in the decision process, and so forth. Therefore it is important for us to stress the methodological differences between the approaches.

Proponents and Opponents.

Let us imagine a discussion where some persons are opponents and others proponents for each methodological approach. We could imagine possible *pro* and *cons* given during such discussion. The authors of the book do not always share the opinions of the proponents and opponents, but they have heard such opinions in different discussions. The collections of the arguments might be interest to the reader.

MAUT

Proponent: The MAUT approach has a strong mathematical base. This approach provides a strong justification of the type of utility function used for aggregation of single-attribute utilities over criteria. Different kinds of independence conditions can be assumed (Keeney and Raiffa, 1976). In the case of criteria dependence, a nonlinear form quite different from a simple additive linear model is chosen.

The involvement of the decision-maker is needed to elaborate a utility function. But after this function is elicited, it is possible to compare many alternatives. Should a new alternative appear, no additional decision-maker efforts are needed. In some difficult cases when general evaluations of alternatives are based on the subjective preferences of several experts (as in the case of plutonium disposition), and the number of alternatives is large, new alternatives could be generated combinations of initial alternatives. Possible inaccuracy in the measurements could be compensated for by sensitivity analysis.

Opponent: The questions posed to decision-makers have no psychological justification. Some questions could be very difficult for humans to understand completely. Decision-makers require special training or orientation in order for MAUT methods to be used.

MAUT is based on a supposition of complete compensation: a deficiency in value for one criterion can be compensated by positive evaluations on other criteria. Otherwise, MAUT cannot give decision-makers the ability to understand the tradeoffs needed to develop a compromise between the criteria.

All preferences are taken at one specific moment, without any checking.

Sensitivity analysis is not a universal tool. Sensitivity could help when the potential dominance of one alternative exists. Otherwise, the possible human errors in evaluation of model parameters are not considered.

AHP

Proponent: AHP is oriented to the real alternatives under consideration, and is very widely used. Comparison of the pairs of alternatives is very easily understood by decision-makers. Comparison of criteria by pairs is the basis for the development of weights; the average "power" of each criterion is expressed through these comparisons.

Let us suppose that there are a small number of alternatives and a relatively small number of criteria with numerical estimates. In such cases the AHP approach focuses directly on the decision goal.

AHP provides a check of the degree that the decision-maker's responses vary from random responses.

Opponent: The formula for aggregation of both criteria and alternative utilities has no formal basis, especially when there is nonlinear utility on criteria scales.

There is no check of criteria independence.

The scale for the transfer from qualitative comparisons to the numbers is different for different decision-makers, and differences could be large. It is impossible to find inevitable human errors in evaluations, except when decision-maker responses are inconsistent.

Special explanation is required for decision-makers to give accurate holistic comparison of criteria importance.

Decision-maker preferences are measured as of one specific time, without possibility for changing assessments through learning.

Outranking Methods

Proponent: Pair-wise alternative comparison is very important to find the degree of compromise needed for selection of the best alternative. The goal of the analysis is to help decision-makers construct a subjective decision rule. The parameters of a model are not very important in the process - but are simply tools in the hands of a consultant.

In a real organization setting, understanding the degree of contradictions between estimations on criteria on the set of given alternatives is required. Veto thresholds and incomparability conditions prevent compensation for the lack of some quality of an alternative (bad estimation on one criterion) by an excessively strong quality (good evaluation on other criteria). The adaptation to the group choice is easy in the ELECTRE method if the weights are nominated by voting of active group members (Chapter 6). Outranking methods do not need sensitivity checks, because they focus on the learning features of multicriteria analysis through constructing outranking relationships based on substantive differences among alternatives.

Opponent: There is no formal justification of the approach. There is no test of possible criteria dependence. The possible errors in decision-maker estimations could not be detected.

Nomination of criteria weights done in a questionable manner could influence a decision.

The concordance, discordance, and veto thresholds required are often not well-understood by decision-makers. Yet outranking relationships are quite sensitive to these selected parameter values.

In the case of possible intransitivity, there is no reasonable base for decision-makers to select one or several alternatives from a cycle of outranking relationships.

VDA

Proponent: The approach has psychological and mathematical basis. In all stages of the method decision-maker and expert interaction uses natural language.

Criteria independence is checked. If criteria are dependent, a different procedure is used. The verbal description of a problem is transformed to obtain independence (Larichev and Moshkovich, 1997).

Measurement of estimates of alternative performance on criteria is subjective process. Therefore it is necessary to consider the precision of the "measurement device". The construction of a decision rule is also a subjective process. Thus the possibilities and limitations of the human information processing system must be considered. All questions of the decision-maker and of experts must be adapted to these limits. People make errors, so there should be procedures for finding possible errors and contradictions. The VDA approach has special procedures for the check of information provided by the decision-maker for contradictions (ZAPROS and ORCLASS methods).

The decision-maker could use the procedure to learn of tradeoffs among alternatives. No special training or orientation is needed for utilization of VDA methods. If there is a great deal of uncertainty in estimations, comparative evaluations could be used (Chapter 7).

Opponent: There are some cases when incomparability (due to lack of reliable information) does not allow identification of the best alternative. In other words, the decision rule might not be decisive enough in cases when a decision must be reached quickly. There is no guarantee that experts could find a better alternative after formulation of directions for improvements of existing alternatives.

Primary Requirements of a Normative Method

Despite the large number and diversity of normative multicriteria methods it is possible to formulate three primary methodological requirements for method results. These requirements reflect the desire of a user to pass on some of the properties of an acceptable tool.

1)   Non-contradictory output from the method: Results should be transitive over the set of alternatives
2)   Method output provides sufficient information to make a decision: The relations between multicriteria alternatives should be complete. After spending a great deal of time and effort in construction decision rules, decision makers want to receive a definite answer – which alternative is best and/or how to rank alternatives based on their general qualities.
3)   Low sensitivity to human error: Each decision method uses some decision maker information to evaluate alternatives. The decision maker has no ideal measuring device such as an oracle. The decision maker can make errors, and demonstrate some contradictions and bias. Therefore, methods must be insensitive to possible errors in the process of information elicitation from the decision maker.

Let us review the extent to which the decision methods we have presented above satisfy these requirements.

Completeness of Relations over Alternatives

The methods based on multiple attribute utility theory and analytic hierarchy process always provide a complete set of relations over alternatives. The output of these methods assign a utility of value to each alternative, which can be used as the basis for ranking. On the other hand, methods based on outranking or verbal decision analysis in general cannot provide this completeness because of the possibility of incomparability between some alternatives. The reason for incomparability is different between these methods. The utilization of two indices for alternative comparison is reason outranking methods may have incomparability. The utilization of only qualitative decision maker evaluations during information elicitation is the reason that verbal decision analysis methods may encounter incomparability. In either case, these latter two methods may not reach a clear decision.

Transitivity and Sensitivity to Human Error

Formally, methods based on multiattribute utility theory and analytic hierarchy process provide transitive results. On the other hand there is a problem with the transitivity for outranking methods. The output of verbal decision analysis methods is always transitive. However, the analysis becomes more complex when one considers the sensitivity of methods to error in human input.

Suppose that during information elicitation from a decision maker we find deviation from the results that are also acceptable to the decision maker. In case of a quantitative parameter measurement, this deviation might be close to the initial value (equivalent lottery value, quantitative criterion weights, etc.). In case of transformation of a qualitative value into a quantitative value directly or through a membership function, we might also discover multiple quantitative values that are acceptable to the decision maker. Overall, we might discover "indifference regions," intervals of values of quantitative parameters, over which decision maker preferences do not vary.

In the framework of multiattribute utility theory and analytic hierarchy process, an implicit supposition is made that the decision maker provides exact and error-free inputs during the process of information elicitation. In other words, we replace a human being by an "oracle" having all numbers in memory and waiting for elicitation of these numbers.

This supposition is far from reality. Using the concept of indifference regions it is easy to show that sensitivity to human errors could lead to intransitivity.

Let us take two different values in an indifference region of a parameter $X: x_1$ and x_2. Suppose that by x_1 alternatives A, B, and C have the following utilities:

$U(A) = u + \varepsilon;\ U(B) = u;\ U(C) = u - \varepsilon.$

Where ε is a small quantitative parameter.

Then, $A \Rightarrow B \Rightarrow C$.

Let us take a different value for the same parameter: x_2. It is possible in a general case to find different alternative utilities:

$U(A) = u - \varepsilon;\ U(B) = u;\ U(C) = u + \varepsilon.$

Then, $C \Rightarrow B \Rightarrow A$.

According to the definition of an indifference region, decision maker preferences over this region are the same. Combining the two alternative orders obtained above, we have:

$A \Rightarrow B \Rightarrow C \Rightarrow B \Rightarrow A.$

Thus, we obtain an intransitivity relation over the alternative set. We see that the last requirements for evaluation of a decision method's output are dependent. For outranking methods, the necessary condition for transitivity is given in Anich and Larichev (1996). For verbal decision analysis methods the output is always transitive and checked to assure that there are no contradictions.

As a result of the above analysis, we can make the following conclusion:

There are no multicriteria decision methods satisfying all three requirements. The methods based on multiattribute utility theory and analytic hierarchy process could provide complete relations over the alternative set. But they are sensitive to human errors, and due to the property of transitivity, will not always provide the correct result. Methods based on outranking relations are less sensitive to human input error. But they cannot guarantee completeness of relations over the set of alternatives. Methods based on verbal decision analysis are not sensitive to human input errors due to the utilization of much more reliable qualitative operations during decision maker elicitation. However, verbal decision analysis methods cannot guarantee complete relations over the alternative set. No normative multicriteria method is perfect and free from deficiencies.

Conclusion

The discussions on merits and deficiencies of different approaches demonstrate that there is no approach best for all cases. A consultant must have knowledge of all approaches and be able to apply the most appropriate method depending on the particular features of the specific decision problem. The cases given in the book were presented in an effort to represent the great diversity of real life siting problems.

The problem of selection of a decision method for a practical problem has a cultural background. The existence of different schools in decision research defines to a great degree the toolbox and professional habits of a specific consultant. There are positive and negative consequences of this situation.

Positively, it is possible to stress the professional habits developed by the consultant in dealing with one kind of analytic technique. Decision analysis is a combination of art and science. To develop the art of successful application of a decision analytic technique it must be practiced.

Negative consequences of consultant attachment to a specific analytic technique include limiting the number of possible analytic tools that could be successfully used for different kinds of practical problems, and utilization of the technique in cases when a different approach could be more efficient. To a large degree, the negative consequences are defined by the state-of-the-art in decision analysis. This field of research lacks sufficient independent comparative studies of the conditions for application of different analytic techniques in practical tasks. We earlier stressed the difficulties in undertaking such studies. But the maturity of research field could be defined by its ability to classify tools and tasks. On the other hand, the reason for this situation is lack of information about the different approaches. The number of publications comparing different approaches in a sound manner is limited.

The practical cases presented in this book demonstrate that multicriteria decision making (MCDM) is a multidisciplinary field. The development of decision methods requires knowledge of diverse fields such as mathematics, cognitive psychology, and organizational behavior. The discussion presented above demonstrates that a new approach to MCDM is needed.

Essentially MCDM is applied science. The primary goal of research in MCDM is to develop tools to help people to make more reasonable decisions. But sometimes the development of such tools is connected with problems in applied mathematics, cognitive psychology, and organizational behavior. Because MCDM is an applied field, practitioners play an important role. They must synthesize results from different disciplines, and only they can find new theoretical problem statements for researchers in the other disciplines. The quality of their practical work depends a great deal on interdisciplinary knowledge as well as on the personal skill of consultants.

Questions important for the future of MCDM include how it should be taught. If one looks at MCDM as a branch of applied mathematics, it could be taught in general coursework of operations research. But if MCDM is multidisciplinary, it must be taught in a quite different manner. Along with utility theory, multiattribute utility theory, prospect theory, knowledge about human memory, organization of human information processing systems, and human behavior in organizations must be studied. Such courses of lectures could give a better picture of MCDM, and could show to students the real face of decision-making involving multiple criteria. The number of publications comparing different approaches in a sound manner is very limited.

References

Anich, I. And Larichev, O. The method ELECTRE and problem of cycles on alternatives set. *Automation and Control* 1996, 108-118.

Barzilai, J., Cook. W. and Golany, B. The analytic hierarchy process: Structure of the problem and its solutions. In *Extremal Methods and Systems Analysis II*, A. Ben-Israel, A. Ben-Tal, B. Golany, K.O. Kortanek and J.J. Rosseau, eds., Westport, CT: Greenwood Press, 1990.

Brans, J.P. and Vincke, P. A preference ranking organization method: The PROMETHEE method, *Management Science* 31, 1985, 647-656.

Budescu, D. and Wallsten, T. Processing linguistic probabilities: General principles and empirical evidence, *The Psychology of Learning and Motivation* 32, Academic Press, 1995.

Dyer, J.S. and Sarin, R.K. Measurable multiattribute value functions, *Operations Research* 27, 1979, 810-822.

Edwards, W. and Barron, F.H. SMARTS and SMARTER: Improved simple methods for multiattribute utility measurement. *Organizational Behavior and Human Decision Processes* 60, 1994, 306-325.

Erev, I and Cohen, B. Verbal versus numerical probabilities: Efficiency, biases, and the preference paradox, *Organizational Behavior and Human Decision Processes* 45, 1990, 1-18.

Fischoff, B. The real world: What good it it? *Organizational Behavior and Human Decision Processes* 65:3, 1996, 232-248.

Flanders, N.E., Brown, R.V., Andre'eva, Y. and Larichev, O. Justifying public decisions in Arctic oil and gas development: American and Russian approaches. *Arctic* 51:3, 1998, 262-279.

Gonzalez-Vallejo, C. and Wallsten, T. The effects of communication mode on preference reversal and decision quality, *Journal of Experimental Psychology: Learning, Memory and Cognition* 18, 1992, 855-864.

Huber, B. and Huber, O. Development of the concept of comparative subjective probability. *Journal of Experimental Child Psychology* 44, 1987, 304-316.

Keeney, R.L.   *Value-Focused Thinking*, Cambridge, MA: Harvard University Press, 1992.

Keeney, R.L. and Nair, K. Evaluating potential nuclear power plant sites in the Pacific northwest using decision analysis, IIASA Professional Paper no. 76-1.   Also in Bell, D.E., Keeney, R.L. and Raiffa, H., eds., *Conflicting Objectives in Decisions*, chapter 14, New York: Wiley, 1977; and in Keeney, R.L., *Siting Energy Facilities*, Chapter 3, New York: Wiley, 1980.

Keeney, R.L. and Raiffa, H. *Decisions with Multiple Objectives: Preferences and Value Tradeoffs*, New York: Wiley, 1976.

Larichev, O.I. Cognitive validity in design of decision-aiding techniques, *Journal of Multicriteria Decision Analysis* 1:3, 1992, 127-138.

Larichev, O.I. and Moshkovich, H.M.   *Verbal Decision Analysis for Unstructured Problems*, Boston: Kluwer Academic Press, 1997.

Larichev, O.I., Olson, D.L., Moshkovich, H.M. and Mechitov, A.I. Numerical vs. cardinal measurements in multiattribute decision making: How exact is exact enough? *Organizational Behavior and Human Decision Processes* 64:1, 1995, 9-21.

Lootsma, F.A. Scale sensitivity in a multiplicative variant of the AHP and SMART, *Journal of Multi-Criteria Decision Analysis* 2:1, 1993, 87-110.

Pitz, G.F. *DECAID Computer Program*, Carbondale, IL: University of Southern Illinois, 1987.

Raiffa, H. *Decision Analysis: Readings*, Boston: Addison-Wesley, 1968.

Roy, B. Classement et choix en présence de points du vue multiples (la méthode ELECTRE), *Rev. Francaise Automat., Informat., Recherche Opérationelle* 8, 1968 (in French).

Roy, B. *Methodologie Multicritere d'Aide a la Decision*, Paris: Economica, 1985 (in French).

Roy, B. *Multicriteria Methodology for Decision Aiding*, Dordrecht: Kluwer Academic Publisher, 1996.

Roy, B. and Bouyssou, D. Comparison of two decision-aid models applied to a nuclear power plant siting example, *European Journal of Operational Research* 25, 1986, 200-215.

Saaty, T.L. *The Analytic Hierarchy Process*, New York: McGraw-Hill, 1980.

Salo, A. and Hämäläinen, R.P. Preference programming through approximate ratio comparisons. *European Journal of Operational Research* 82:3, 1995, 458-475.

Smith, G.R. and Speiser, F. *Logical Decision: Multi-Measure Decision Analysis Software*, Golden Co: PDQ Printing, 1991.

9 SUPPORT TO THE MULTIATTRIBUTE DECISION PROCESS

Multicriteria siting choice and particularly siting waste disposal facilities are very difficult decisions. This book reviewed some of the approaches used across the world to help make these difficult decisions on the basis of more thorough consideration of inevitably conflicting criteria and of inevitably conflicting groups. Four representative applications were presented, and the decision processes as well as group interactions involved were discussed.

Cost-benefit analysis is often used for this class of problem, but requires monetary estimations of subjective factors, and requires estimation of long-term effects, both sources of high degrees of error. Multiattribute utility theory provides a systematic means of estimating tradeoffs. If sufficient time and data is available, this is a sound and widely accepted means of analyzing waste siting problems. If higher levels of uncertainty and subjectivity exist, decision-maker learning about tradeoffs can be obtained through outranking methods. If the degree of measurement difficulty is higher yet, verbal decision analysis provides a sound method of analysis for initial selection and improved design of alternatives.

The preceding chapter reviewed technical differences in methods, and how these differences relate to supporting decision making. This concluding chapter discusses the new features of multiattribute location decision as a special class of decision making problems: unique mixture of individual and collective choice, new role of experts, generation of improved alternatives. The role of different groups in the decision processes, and the implications of the multiple attribute multiple actor model are considered. We shall give short view of existing and future technical tools for siting decisions and the potential of information technology.

The Mixture of Individual and Collective Choice in Multicriteria Location Decisions.

According to widely accepted classification the siting decision could be individual or collective. The cases of multicriteria location decisions presented above allow us to say that this classification is too simple. The procedures of real location decisions represent the complex mixture of individual and collective choice.

Usually in the location decision there is a central figure-head of governmental or privet organization having the primary responsibility for the decision. The policy and preferences of this organization play am important role in the process of the choice. But the active groups also are very influential.

Not only is it important for analyses to reflect a variety of judgment criteria, it is also important to consider those who have a say in the decision. In the Finnish waste disposal decision process the opinions of local citizens in the location decision were elicited as an integral part of the process. This input was used in the form of voting to obtain weights for the outranking analysis.

Consideration of public opinion is very important in U.S. siting decisions as well. It is implemented as part of the decision process, in the form of representational approval from Congress, as well as part of the design process. The location of a nuclear waste facility discussed in Chapter 3 currently is a critically important social issue. Nuclear waste is building up at over 70 reactor sites in the United States (Schuler, 1999). There is strong pressure from these plants, which have collected a surcharge from consumers for years to pay for permanent storage, for the federal U.S. Government to take the waste material. This is causing problems by raising the cost of electricity. The State of Nevada, where the Yucca Mountain site planned for permanent storage is located, adamantly opposes its use (Preston, 1999). The federal government is carefully proceeding with studies, along with test development (Armistead, 1998). The federal government reached a decision as outlined in Chapter 3. Currently, use of the Yucca Mountain facility has been delayed judicially until 2007. If approved, the site will not be available until 2010 anyway (Schuler, 1999). It remains to be seen whether or not it will actually be used.

The location of pipeline facilities in the Soviet Union as well as in the country of Russia involved decision making by government agencies representing the interests of citizens as outlined in Chapter 7. In the former Soviet Union as well as in Russia it is necessary to obtain the agreement of local administration for the construction of technical objects. Russian citizens play an important role in the decision making process. Recently they protested against some variants of the options under consideration. In the case of Yamal such protest by the local population and influential environmental groups brought pressure on the organizations responsible for final choice. As the result the decision has been postponed.

Active groups are becoming more and more influential. This means that constant struggle for the power is going on in the process of decision making. For the organization responsible for the decision, it is necessary to convince the active groups that the chosen variant satisfies their preferences. Very often side-by payments are made in the form of additional costs due to modifications of chosen variant or through compensation for local citizens. In the cases presented in different chapters the relationship of influence and power between decision making organizations and active groups was clearly important.

In the case of AHP application for water resource planning the decision apparently was acceptable as for decision-making organization as well as for active groups (Chapter 5).

In the Finnish waste disposal decision process the active groups were satisfied by direct influence on criteria weights. They accepted as given the general method of alternative evaluation –ELECTRE approach (Chapter 6). In the choice of gas pipeline route location in Soviet Union, the preferences of active parties defined the choice of a new, improved variant in the decision. The increased cost of this variant represents the payment for the agreement between the active parties and the governmental organization responsible for the choice (Chapter 7).

For three other cases: siting a high-level nuclear waste repository (Chapter 3), disposing of excess plutonium (Chapter 4), choice of gas pipeline route in Yamal peninsula (Chapter 7) the relation of power between active groups and decision making organization was quite different.

The State of Nevada, where the Yucca Mountain site planned for permanent storage of nuclear waste is located, opposes its use (Preston, 1999). The strong opposite reaction of local administration was also in the case of plutonium long-term disposition. In the case of the Yamal gas pipeline, the local administration and population was strongly opposed to the land variant. The decisions were postponed for the long time.

It seems clear that when a consensus is achieved between majority of population (or active groups) and the decision making organization, this does not mean that a final decision has been reached. Even though a clear consensus exists in the high-level waste problem (Yucca Mountain), an important minority constituency adamantly opposes the idea. The ability of an active minority to stop a decision is now the feature of decision process in democratic countries.

That is why, the idea of resolving multiattribute group decision problems through voting is not sufficient to reach problem solution. Voting is in fact the method officially available for problem resolution. Representative voting has been applied in some cases, but the process has not yielded a clear decision, probably because forcing a decision on unwilling citizens is not politically acceptable to the voting representative.

Multiattribute analysis does not guarantee reaching a decision. There continue to be serious debates about location of nuclear waste disposal sites in the United States, years into the process, and well into the period when the facility is desperately needed. Consensus clearly will not be reached. Voting is the official mechanism of resolution, but that process is currently insufficient to reach a final conclusion as well. Multiattribute utility analysis was applied, clearly uncovering the tradeoffs involved among important criteria. While we would point to this analysis as a resounding success, it must be recognized that multiattribute analyses are not guarantors of reaching decisions. They play a valuable role in demonstrating tradeoffs. Humans are still responsible for the final decisions.

Unfortunately, for the decisions vitally important for many active groups the stalemate could be often the output of decision process. Several examples of such stalemate are given in the book of Kunreuther et al.(1983).

We have here the paradox typical for the collective decisions (Arrow, 1959; Sen, 1977; Sen, 1997): a decision rule can not be simultaneously *democratic, decisive and rational.*

The Role of Experts in Siting Decisions

The typical role of experts in the decision problems is to provide the information needed for the evaluation of the factors important for evaluation of alternatives.

In multiattribute siting decision sometimes experts play this role. But there are difficult problems where expert knowledge is limited.

For example, the existing knowledge does not allow to estimate in reliable way the consequences of long-term plutonium disposal in the places different by geological conditions. Different experts currently give different estimations for the reliability of Yucca Mountain site for nuclear waste facility. Different experts give different estimations for the reliability of seashore geology on the Yamal peninsula.

This implies that human knowledge is limited for long-term strategic decisions. In spite of development of advance mathematical models and results of long-term nature observations some evaluations are very approximate. In some cases relative evaluations could be more reliable than absolute evaluations (Chapter 7).

More important is that the active parties and public do not always believe in expert estimations. For example, an ordinary citizen from does not believe exact numerical estimation of factors over a period of the next 300 years. The disputes between different experts increase the lack of confidence in expert estimations.

Thus strategic siting problems place a special requirement on experts. The experts are to define as clearly as possible the existing limits of knowledge in the problems under consideration. It must be clear that for the decision-maker and for active groups it is better to have an approximate but reliable estimate instead of an exact and unreliable estimate. The role of expert is now to explain to general public the logic and justification of estimations in an understandable way given different factors. Where it is possible it is better to give a suggestion how a source of uncertainty can be avoided instead of giving uncertain estimates of expected outcomes. The payment for the elimination of uncertainty sources could be expensive but the only way for mankind to use unknown and uncertain technologies.

Generating Improved Alternatives

Traditionally, multiattribute methods have focused on evaluating a set of given alternatives. Recently, more emphasis has been placed on designing improved alternatives in the context of selection decisions (Keeney, 1992). The idea of generating improved multiattribute alternatives has also been proposed in the context of mathematical programming (Olson, 1995).

There is ample evidence of this process at work in the applications we have reviewed. In the case of disposal of weapons-grade plutonium, a two-phase decision process was applied. In phase one, a large set of diverse options were generated, and a process of evaluation relative to meeting minimum-acceptable criteria was used to reduce this large set to about a dozen variants of the three basic options of vitrification, reactors, and boreholes. In phase two, experts from national laboratories generated improved alternatives, increasing the number of options considered to about eighteen. This decision process clearly involved a workable means of applying expertise to generate improved solutions.

Four of the governmental decision processes we examined demonstrate the potential of analyses to generate improved decisions. The pipeline decisions in Georgia, in the Yamal Peninsula, and in Niakuk, Alaska all involved opportunities to improve the initial set of alternatives. In the Georgia case, the original alternatives were unacceptable, so an improved solution was generated. The Yamal Peninsula and Niakuk cases were used in Chapter 7 of this book to demonstrate generating improved solutions through the ASTRIDA method (Berkeley, et al., 1991). In the Niakuk case, British Petroleum actually adopted an improved solution, at the suggestion of the Corps of Engineers. In the weapons-grade plutonium disposition application, developing improved solutions was part of the planned decision process (see Dyer, et al., 1998 for details).

Existing Multiattribute Tools for Choices

The process of decision making needs to consider not only multiple attributes, but also the decision making process that clearly appears in the MAMP models. The rationalist philosophy behind multiattribute utility theory is based on an atomist view, where overall value is equal to a function considering the component contributions of an alternative on each of the criteria (a compensatory model). This model allows for nonlinear relationships in single attribute utility functions, which are often used. It also allows for nonlinear aggregation functions, which usually are not used, especially in public decision making. At an extreme, this approach could easily convert into cost-benefit analysis. (Using cost as the base comparison for all lottery tradeoff questions would accomplish that.) While analytic hierarchy process is different than MAUT, the underlying value function is based on the simplest version of the MAUT idea - value is the sum of the contributions of each criterion.

European approaches differ. Both outranking and verbal decision analysis models allow for decision-makers to obtain a partial order of alternatives that identify those alternatives that cannot be established as inferior. The outranking models provide a means for decision-makers to construct value functions using a variety of measurement forms. Verbal decision analysis gives decision-makers a means to base preference statements on sounder input.

Technical Support Available for Location Decisions

Decision making for siting decisions in the Twentieth Century required decision-makers to undertake the following steps:

1) collection of information about different locations with estimates on multiple criteria (cost, influence on environment, personal, participation of active groups and so on).
2) Preliminary selection of a site, detailed investigation, and negotiation with active groups.
3) Final decision.

We could guess that in the Twenty-First Century siting decisions will be accomplished in a different manner. The information base for such decisions will be different. The fast development of geographic information systems (GIS) creates new possibilities for expanding the data available for siting decisions. Huge amounts of information can be stored in a GIS.

Using the Internet, decision-makers can receive all of the information needed for the first step of a decision with much better quality than has been the case in the past. For example, local authorities looking for an investor could develop a special map showing prospects and limitations for different geographic locations in a specified area. The first steps in these developments are occurring now in a number of countries.

Summary

This book has shown how major multiple criteria decision analysis methods have been applied in waste siting decisions. The multiple attribute multiple actor model demonstrates the dynamic aspects of this class of decision. Chapter 8 provided a comparison of the technical operation of these methods. Different models based upon different philosophies are available. This chapter has discussed some final implementation issues. An important aspect of this class of decision problem is the generation of better alternatives. The skill of a facilitator helping to the decision making organization to find a compromise with active groups became more and more important.

We could say that the information support of DSS in the next century will be more advanced. But the crucial problem of decision making under conditions of uncertainty and multiple criteria will remain the same. Advanced DSS can help a DM, but cannot replace him or her. The ability of human beings to take necessary risks and to use intuition in place of missing objective data are the most important ingredients of successful decision processes. That is why decision methods and DSS will be needed in the next century for successfully making important decisions.

The practical cases presented in this book demonstrate that multicriteria decision-making (MCDM) is a multidisciplinary field. The development of decision methods requires knowledge of diverse fields such as mathematics, cognitive psychology, and organizational behavior. The discussion presented in Chapter 8 demonstrates that a new approach to MCDM is needed.

Essentially MCDM is applied science. The primary goal of research in MCDM is to develop tools to help people to make more reasonable decisions. But sometimes the development of such tools is connected with problems in applied mathematics, cognitive psychology, and organizational behavior. Because MCDM is an applied field, practitioners play an important role. They must synthesize results from different disciplines, and only they can find new theoretical problem statements for researchers in the other disciplines. The quality of their practical work depends a great deal on interdisciplinary knowledge as well as on the personal skill of consultants.

Questions important for the future of MCDM include how it should be taught. If one looks at MCDM as a branch of applied mathematics, it could be taught in general coursework of operations research. But if MCDM is multidisciplinary, it must be taught in a quite different manner. Along with utility theory, multiattribute utility theory, prospect theory, knowledge about human memory, organization of human information processing systems, and human behavior in organizations must be studied. Courses of lectures of this type could give a better picture of MCDM, and could show to students the real face of decision-making involving multiple criteria.

References

Armistead, T.E., Yucca Mountain boring begins. *Engineering News Record*, vol. 240, no. 16, 1998, 16.

Arrow, K.J., Rational choice functions and orderings. *Economica*, vol. 26, 1959, 121-127.

Berkeley, D., Humphreys, P., Larichev, O., and Moshkovich, H. Aiding strategic decision making: Derivation and development of ASTRIDA. In Y. Vecsenyi and H. Sol, eds., *Environment for Supporting Decision Processes*, North-Holland, Amsterdam, 1991.

Dyer, J., Edmunds, T., Butler, J.C., and Jia, J. A multiattribute utility analysis of alternatives for the disposition of surplus weapons-grade plutonium. *Operations Research* vol. 46, no. 6, 1998, 749-762.

Keeney, R.L. *Value-Focused Thinking: A Path to Creative Decisionmaking*, Cambridge, MA: Harvard University Press, 1992.

Kunreuther, H.C., Linnerooth, J., Lathrop, J., Atz, H., Macgill, S., Mandl, C., Schwarz, M. and Thompson, M. *Risk Analysis and Decision Processes: The Siting of Liquified Energy Gas Facilities in Four Countries*. Springer-Verlag, Berlin, 1983.

Olson, D.L. Multiobjective programming as a means of generating decisions. *Research in Public Policy Analysis and Management*, JAI Press, vol., 6, 1995, 273-290.

Preston, M., Locals disagree on nuclear dump. *American City & County*, vol., 114, no. 2, 1999, 12.

Schuler, J.F., Jr., Nuclear waste's slow boil. *Public Utilities Fortnightly*, vol. 137, no. 11, 1999, 34-37.

Sen, A., Social choice theory: A re-examination. *Econometrica*, vol. 45, no. 1, 1977, 53-89.

Sen, A. *Choice, Welfare, and Measurement*, Cambridge, MA: Harvard University Press, 1997.

AUTHOR INDEX

SUBJECT INDEX